Now try this...

▶ A PROGRESSIVE MATHEMATICS COURSE FOR STUDENTS

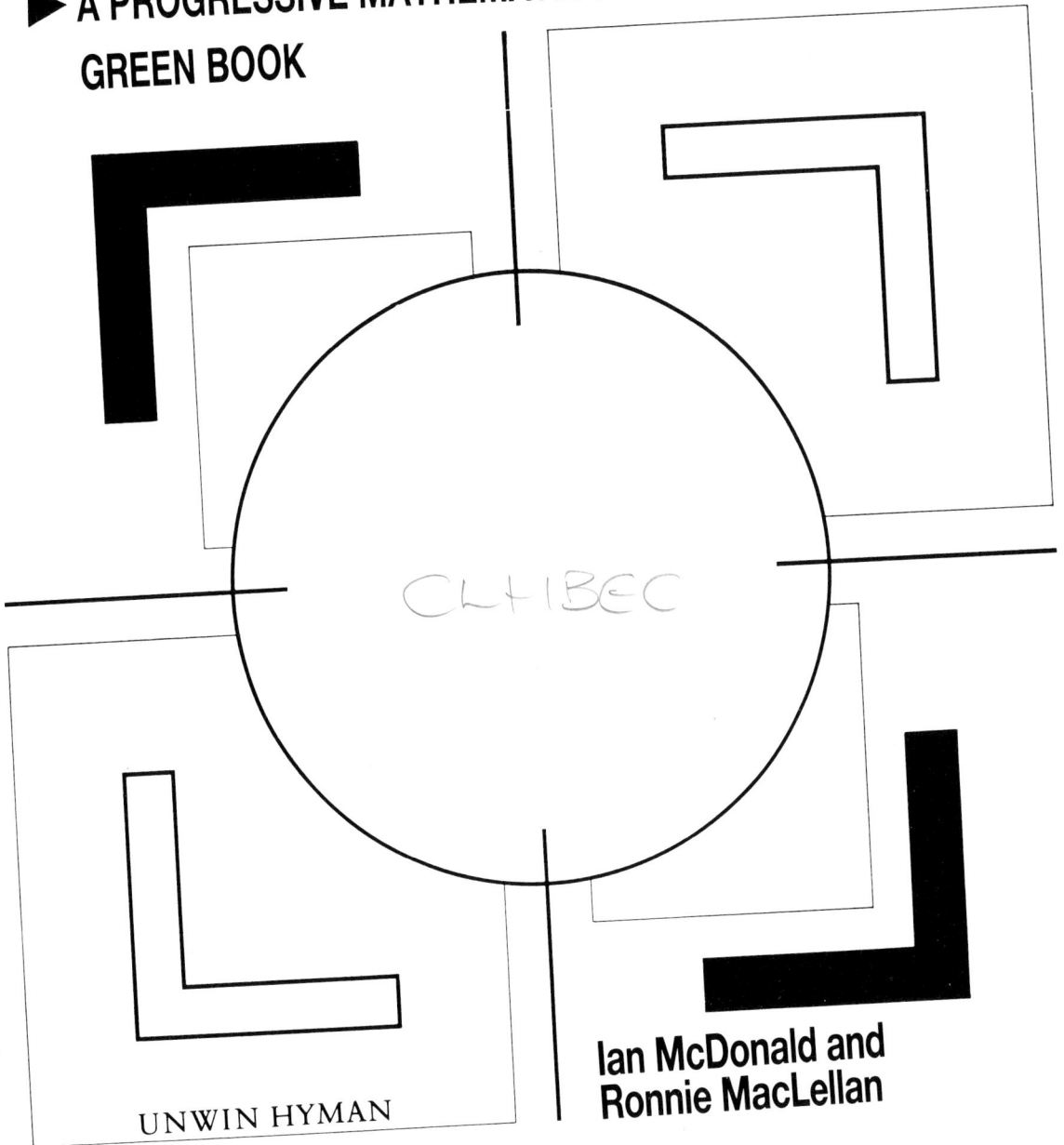

GREEN BOOK

CLHBEC

Ian McDonald and
Ronnie MacLellan

UNWIN HYMAN

ACKNOWLEDGEMENTS

The authors and publishers would like to thank the following for their permission to reproduce the photographs and illustrations that appear on the pages indicated:

Alan International Hairdressing Schools, London, 15, 38
Barnaby's Picture Library, 156
British Architectural Library, RIBA, London, 11
John Cleare, Mountain Camera, 121
Sydney Harris, 44
Adam Hart-Davis, cover
Meat and Livestock Commission, 41
Munro and Forster, Public Relations Ltd, 44
NASA, 45, 57
Olympus Optical Co. (UK) Ltd, 105, 120
Alan Pascoe Associates Ltd, 105
John Reader, Field Studies Council, 105
Royal Observatory, Edinburgh, 49
Shell Photographic Library, 24
Eric Thorburn, 14

Published in 1990 by
UNWIN HYMAN LIMITED
15–17 Broadwick Street
London W1V 1FP

© I McDonald and R MacLellan 1990

Designed by Pete Lawrence

British Library Cataloguing in Publication Data
McDonald, Ian
    Now try this: a progressive mathematics course for students.
    1. Mathematics
    I. Title II. MacLellan, Ronnie
    510

ISBN 0444 8025 3

Artwork by RDL Artset Ltd
Typeset in Hong Kong by Colorcraft Ltd
Printed and bound by New Interlitho, Italy

INTRODUCTION

WHY TRY THIS?

Some books are written for teachers—this book is written
for students.

We are sure that you will be able to work through most of
it by yourself. There are, however, some tricky parts
where your teacher's help will be necessary.

You will find that a unit contains our worked examples
and exercises for you to do (*Try These*). In this way, your
maths skills should be developed so that, at the end of the
unit, you can tackle a real investigation (*Now Try This*).

At the back of the book, there are answers to all the
exercises. These allow you to check your progress. At the
end of each unit are a number of *Key Questions* with no
answers given. These allow your teacher to monitor your
progress.

Each unit has a *Revision Chapter*. This allows you to
revise your skills from time to time and especially before
important examinations.

Maths is thought to be important—that is why you are
studying it. We have tried to show applications of the
maths throughout the book e.g. in the unit on
simultaneous equations, we show how linear programming
can help in running a small business; navigational
problems are tackled in the trigonometry unit; and in the
last unit, we use statistics to compare the examination
results of the two classes.

We believe you will enjoy using the book, but don't take
our word for it—

TRY IT!

Best of luck in your exams—Standard Grade or GCSE.

Ronnie MacLellan and Ian McDonald

CONTENTS

1 FACTORS 1
- Common factors 3
- Multiplying 5
- Other factors 7

2 EQUATION OF A STRAIGHT LINE AND INEQUALITIES 14
- Equation of a straight line 16
- Inequations 22
- Solving inequalities 28
- More on solution sets 30

3 SIMULTANEOUS EQUATIONS AND LINEAR PROGRAMMING 33
- Linear programming 36
- Linear programming—practical applications 37

4 INDICES AND ALGEBRAIC FRACTIONS 44
- Rules of indices 46
- Numbers in standard form 48
- Surds 50
- Rationalising the denominator 51
- Algebraic fractions 52
- Multiplication 52
- Division 52
- Simplifying fractions 53
- Addition and subtraction 54
- Equations with algebraic fractions 55

5 QUADRATICS 57
- Iterative (repetitive) methods 63
- of solving quadratic equations
- A formula for solving quadratic equations 65
- Quadratic equations—with fractions 68

6 FUNCTIONS 71
- Intervals on the real number line 73
- Combination function 76
- The exponential function 78
- Inverse functions 80

7 TRIGONOMETRIC FUNCTIONS AND GRAPHS 84
- Combination functions 93
- Composite functions 95
- The tangent function 99

8 PROPORTIONALITY 105
- Proportion—direct and inverse 106
- Direct proportionality (variation) 108
- Arc of a circle 110
- Other examples of direct proportionality 111
- Area of a sector of a circle 114
- Inverse proportionality (variation) 115
- Joint variation 118

9 TRIGONOMETRY 121
- SOH CAH TOA 122
- The sine rule 125
- The cosine rule 130
- Sine rule or cosine rule 132
- The area of a triangle 133
- Trig ratios—in 3-dimensions 135
- Trigonometric equations 136

10 TRANSFORMATIONS AND MATRICES 140
- Transformations and matrices 148
- Combinations of transformations 152

11 STATISTICS 156

REMINDERS AND REVISION 166
FACTORS 166
STRAIGHT LINE AND INEQUALITIES 168
SIMULTANEOUS EQUATIONS AND LINEAR PROGRAMMING 170
INDICES AND ALGEBRAIC FRACTIONS 172
QUADRATICS 174
FUNCTIONS 177
TRIG FUNCTIONS AND GRAPHS 179
PROPORTIONALITY 186
TRIGONOMETRY 188
TRANSFORMATIONS AND MATRICES 191
STATISTICS 194

ANSWERS 197

FACTORS — THE END PRODUCT

After you have worked through this unit, you should be able to understand the following examples.

$$25a^2 - 9b^2 = (5a - 3b)(5a + 3b)$$

A rectangle is
1 unit more in length and
1 unit less in width than
a square.

The area of the square
in square units is x^2.

The area of the rectangle
in square units is
$(x + 1)(x - 1) = x^2 - 1$

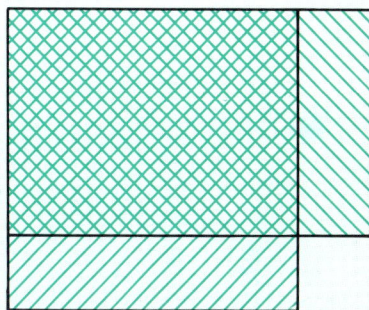

x

The difference in area (square − rectangle) is 1.

$$6y^2 - 23yz + 20z^2 = (2y - 5z)(3y - 4z)$$

$$\frac{1 - 4p^2}{2 - p - 6p^2} = \frac{1 + 2p}{2 + 3p}$$

A rectangle is
3 units more in length and
2 units less in width than
a square.

The area of the square
in square units is x^2.

The area of the rectangle
in square units is
$(x+3)(x-2)=x^2+x-6$.

The difference in area
(square − rectangle) is $6-x$.

$x=5$

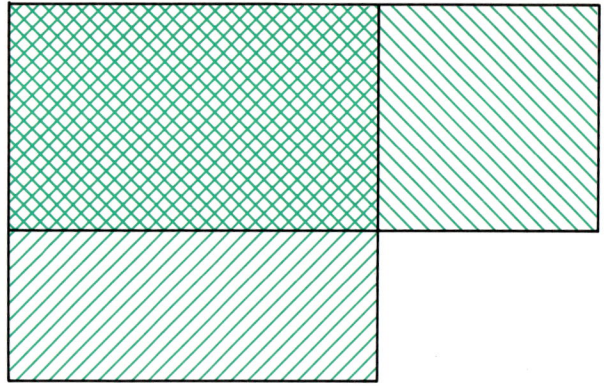

$x=6$

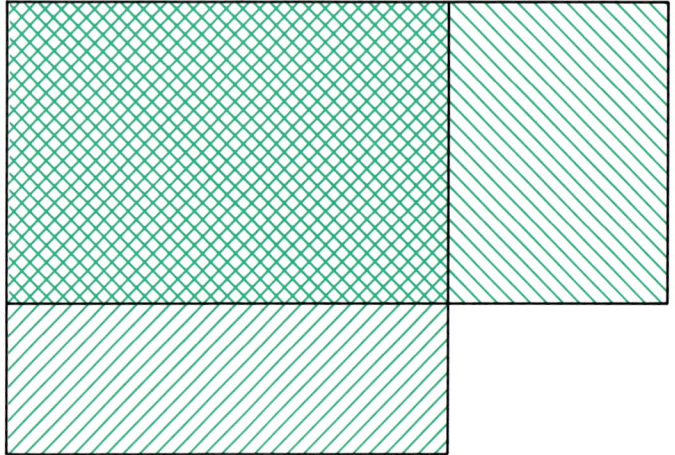

$x=7$

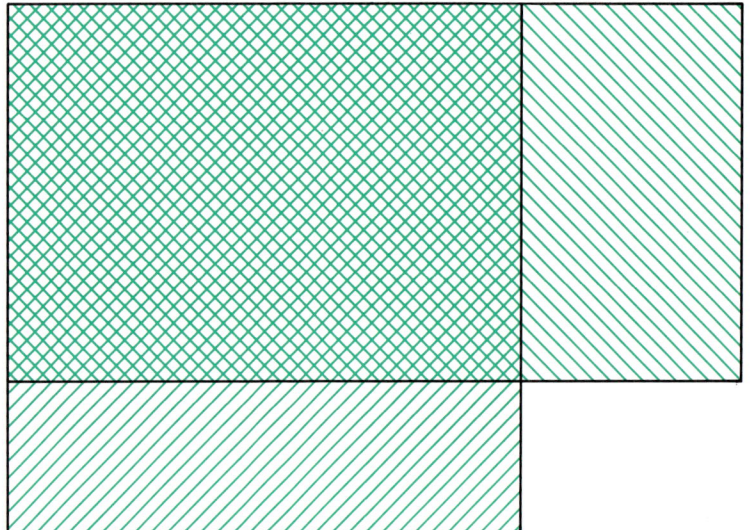

The square has a larger area for values of x less than 6.

The rectangle has a larger area for values of x more than 6.

The areas are equal when $x=6$.

2

Common factors

We saw in an earlier book that:

$2p + 6q$ can be written as $2(p + 3q)$.

Let us look at this more closely.

The **factors** of $2p$ are 2 and p: $2p = 2 \times p$.
The **factors** of $6q$ are 2, 3 and q: $6q = 2 \times 3 \times q$.

$2p$ and $6q$ have a **common factor**, 2.

$2p + 6q = 2 \times p + 2 \times 3 \times q = 2 \times (p + 3 \times q)$.

So, $2p + 6q = 2(p + 3q)$
↑
common factor

Try this

1 (a) $4a + 8b = 4(\ ?\ )$ (c) $4x + 8 =$ (e) $6x - 2y =$
   (b) $6c - 3d =$ (d) $4a + 4 =$ (f) $3b - 6 =$

Here are two more difficult examples.

EXAMPLE

The factors of $5xy$ are 5, x, y: $5xy = 5 \times x \times y$.
The factors of $10yz$ are 2, 5, y, z: $10yz = 2 \times 5 \times y \times z$.

$5xy + 10yz = 5 \times x \times y + 2 \times 5 \times y \times z = 5 \times y(x + 2 \times z)$

The factors of $5xy + 10yz$ are 5, y and $(x + 2z)$.

So, $5xy + 10yz = 5y(x + 2z)$.

EXAMPLE

The factors of p^2q^2r are p, (p), q, (q), r: $p^2q^2r = p \times p \times q \times q \times r$.
The factors of pqr^2 are p, q, r, (r): $pqr^2 = p \times q \times r \times r$.

$p^2q^2r - pqr^2 = p \times p \times q \times q \times r - p \times q \times r \times r = p \times q \times r(p \times q - r)$.

Factors of $p^2q^2r - pqr^2$ are p, q, r and $(pq - r)$.

So, $p^2q^2r - pqr^2 = pqr(pq - r)$.

Try this

2 (a) $2ab + 2bc = 2b(\ ?\ )$ (e) $4xy^2 - 8x^2y =$ (h) $21x^2y - 7xy =$
   (b) $3pq - 6pr =$ (f) $7a^2b - 3ab^2 =$ (i) $x^2y^2z - xy^2z^2 =$
   (c) $9x^2y - 3x^2z =$ (g) $8l^2m^2 + 4lm =$ (j) $l^2m^2n^2 - lmn =$
   (d) $5pqr - 10qr =$

Calculate the value of $\dfrac{5xy + 10yz}{10yz - 5xy}$

when $x = 2$, $y = 3$, $z = 7$

This does not look too easy at first.

Working

$$5xy + 10yz = (5 \times 2 \times 3) + (10 \times 3 \times 7)$$
$$= \quad 30 \quad + \quad 210$$
$$= \quad\quad 240$$
$$10yz - 5xy = (10 \times 3 \times 7) - (5 \times 2 \times 3)$$
$$= \quad 210 \quad - \quad 30$$
$$= \quad\quad 180$$
$$\frac{5xy + 10yz}{10yz - 5xy} = \frac{240}{180}$$
$$= \frac{4}{3}$$

Why not try to find factors before calculating?

Working

$$5xy + 10yz = 5y(x + 2z)$$
$$10yz - 5xy = 5y(2z - x)$$
$$\frac{5xy + 10yz}{10yz - 5xy} = \frac{5y(x + 2z)}{5y(2z - x)}$$
$$= \frac{x + 2z}{2z - x}$$
$$= \frac{2 + 2 \times 7}{2 \times 7 - 2}$$
$$= \frac{16}{12} \qquad = \frac{4}{3}$$

If we find factors first, we may save calculating time.

Here is a similar example.

Find the value of $\dfrac{p^2q^2r - pqr^2}{p^2qr - pq^2r^2}$

when $p = 2$, $q = 1$, $r = 3$.

$$\frac{p^2q^2r - pqr^2}{p^2qr - pq^2r^2} = \frac{pqr(pq - r)}{pqr(p - qr)} = \frac{pq - r}{p - qr}$$

When $p = 2$, $q = 1$, $r = 3$,

$$\frac{pq - r}{p - qr} = \frac{-1}{-1} = 1$$

Try this

3 (a) Find the value of $\dfrac{ab+bc}{ab-bc}$ when

$a=6$, $b=3\cdot7$, $c=2$.

(b) Find the value of $\dfrac{x^2yz-xyz^2}{xy^2z^2-x^2yz}$ when

$x=3$, $y=4$, $z=-1$.

(c) Find the value of $\dfrac{l^2mn^2-lm^2n}{l^2m^2n-lmn^2}$ when

$l=-3$, $m=4$, $n=-2$.

(d) Find the value of $\dfrac{pqr-p^2q^2r^2}{pqr^2-p^2qr}$ when

$p=-1$, $q=-2$, $r=3$.

(e) Find the value of $\dfrac{5x^2y-10xy^2z}{10xy-5xy^2}$ when

$x=3$, $y=-1$, $z=-2$.

(f) Find the value of $\dfrac{7a^2b-21ab^2}{3a^2b-7ab}$ when

$a=4$, $b=1$.

Multiplying

We know that $2x(3x+2y)=6x^2+4xy$
and that $5y(3x+2y)=15yx+10y^2=15xy+10y^2$.

Try this

4 (a) $2x(5x-2y)=$ (b) $3y(5x-2y)=$
  (c) $x(2x+3y)=$ (d) $-2y(2x+3y)=$

How do we calculate $(2+3)(5-2)$?

Method 1	**OR**	**Method 2**
$(2+3)(5-2)$		$(2+3)(5-2)$
$=\ \ 5\ \times\ 3$		$=2(5-2)+3(5-2)$
$=\ \ \ \ \ 15$		$=\ \ 10-4+15-6$
		$=\ \ \ \ \ \ \ 15$

What can we do with $(2x+3y)(5x-2y)$?
We cannot use **Method 1** since $2x+3y$ cannot be simplified.
We can use **Method 2**.

$(2x+3y)(5x-2y)$
$=2x(5x-2y)+3y(5x-2y)$
$=10x^2-4xy+15yx-6y^2$
$=10x^2-4xy+15xy-6y^2$
$=10x^2+11xy-6y^2$

Try this

5 (a) $(2x+3y)(3x+2y)=6x^2+$ (d) $(x-2y)(2x-y)=$
  (b) $(p+2q)(3p-q)=$ (e) $(3a-4b)(4a+3b)=$
  (c) $(4l+2m)(l-3m)=$ (f) $(5c-d)(c-4d)=$

Let us look at some more examples.

EXAMPLE

$$(3x + 2y)(5x - 7y) = 3x(5x - 7y) + 2y(5x - 7y)$$
$$= 3x \times 5x + 3x \times (-7y) + 5x \times 2y + 2y \times (-7y)$$
$$= 15x^2 - 21xy + 10xy - 14y^2$$
$$= 15x^2 - 11xy - 14y^2$$

So, $(3x + 2y)(5x - 7y) = 15x^2 - 11xy - 14y^2$

$-14y^2$ is $[\boxed{2}\,y \times \boxed{-7}\,y]$

$-11xy$ is $[\boxed{3}\,x \times \boxed{-7}\,y + \boxed{5}\,x \times \boxed{2}\,y]$

$15x^2$ is $[\boxed{3}\,x \times \boxed{5}\,x]$

EXAMPLE

$$(a + 2)(a + 5) = [1 \times 1]a^2 + [1 \times 5 + 1 \times 2]a + [2 \times 5]$$
$$= [1]a^2 + [7]a + [10]$$

So, $(a + 2)(a + 5) = a^2 + 7a + 10$

EXAMPLE

$$(m - 3)(m - 4) = [1 \times 1]m^2 + [1 \times (-4) + 1 \times (-3)]m + [(-3) \times (-4)]$$
$$= [1]m^2 + [-7]m + [12]$$

So, $(m - 3)(m - 4) = m^2 - 7m + 12$

Try this

6 (a) $(a + 1)(a + 2)$ (d) $(y - 1)(y - 3)$
   (b) $(b + 5)(b + 3)$ (e) $(r + 1)(r + 6)$
   (c) $(p - 4)(p - 6)$ (f) $(c - 3)(c - 3)$

EXAMPLE

$$(x - y)(x + 2y) = [1 \times 1]x^2 + [1 \times 2 + 1 \times (-1)]xy + [(-1) \times 2]y^2$$
$$= [1]x^2 + [1]xy + [-2]y^2$$

So, $(x - y)(x + 2y) = x^2 + xy - 2y^2$

EXAMPLE

$$(5 - 2n)(2 - 3n) = [5 \times 2] + [5 \times (-3) + 2 \times (-2)]n + [(-2) \times (-3)]n^2$$
$$= 10 + (-19)n + 6n^2$$

So, $(5 - 2n)(2 - 3n) = 10 - 19n + 6n^2$

Try this

7 (a) $(2x + 3y)(x + y) =$ (f) $(1 + a)(1 - 3a) =$
   (b) $(2a - b)(a - b) =$ (g) $(5 - 3x)(4 - 5x) =$
   (c) $(2p - q)(p + 2q) =$ (h) $(a + 1)(b + 1) =$
   (d) $(3l + 2m)(3l - 2m) =$ (i) $(2 - x)(3 - y) =$
   (e) $(3x - 2)(5x - 3) =$

Other factors

EXAMPLE

We know that $(a+2)(a+3)=a^2+5a+6$.

If we reverse this statement:

$a^2+5a+6=(a+2)(a+3)$,

we see that the factors of a^2+5a+6 are $(a+2)$ and $(a+3)$.

Let us now find the factors of $b^2+7b+12$:

$$b^2+7b+12=(b+N_1)(b+N_2)$$
$$=b^2+(N_1+N_2)b+N_1\times N_2$$
$$=b^2 \qquad 7b \qquad 12$$
$$\qquad \qquad \downarrow \qquad \quad \downarrow$$
$$\qquad \quad add\ to\ 7 \qquad multiply\ to\ 12$$

We need to find N_1 and N_2.

$1\times 12=12$ ✓	$2\times 6=12$ ✓	$3\times 4=12$ ✓
$1+12=13$ ✗	$2+6=8$ ✗	$3+4=7$ ✓

So, $b^2+7b+12=(b+3)(b+4)$.

The factors of $b^2+7b+12$ are $(b+3)$ and $(b+4)$.

Try this

8 (a) $x^2 + 5x + 4 = (x + ?)(x + ?)$ (d) $p^2 + 10p + 21 =$
   (b) $a^2 + 3a + 2 =$ (e) $q^2 + 4q + 3 =$
   (c) $b^2 + 6b + 8 =$ (f) $x^2 + 2x + 1 =$

EXAMPLE

We know that $(m - 3)(m - 4) = m^2 - 7m + 12$

Reversing gives $m^2 - 7m + 12 = (m - 3)(m - 4)$,
and we see that the factors of $m^2 - 7m + 12$ are
$(m - 3)(m - 4)$

EXAMPLE

Let us now find the factors of $m^2 - 6m + 8$.

$$m^2 - 6m + 8 = (m + N_1)(m + N_2)$$
$$= m^2 + (N_1 + N_2)m + N_1 \times N_2$$
$$= m^2 + \quad -6m \quad + \quad 8$$
$$\qquad\qquad\qquad \downarrow \qquad\qquad \downarrow$$
$$\qquad\qquad add\ to\ -6 \qquad multiply\ to\ 8$$

$1 \times 8 = 8$ ✓	$-1 \times (-8) = 8$ ✓	$-2 \times (-4) = 8$ ✓
$1 + 8 = 9$ ✗	$(-1) + (-8) = -9$ ✗	$-2 + (-4) = -6$ ✓

So, $m^2 - 6m + 8 = [m + (-2)][m + (-4)]$
$$= (m - 2)(m - 4)$$

The factors of $m^2 - 6m + 8$ are $(m - 2)$ and $(m - 4)$.

Try this

9 (a) $x^2 - 5x + 6 = (x - ?)(x - ?)$ (d) $x^2 - 6x + 5 =$ (g) $m^2 + 5m + 4 =$
   (b) $p^2 - 6p + 8 =$ (e) $q^2 - 10q + 21 =$ (h) $y^2 + 7y + 10 =$
   (c) $p^2 - 6p + 9 =$ (f) $q^2 - 10q + 16 =$ (i) $a^2 + 4a + 3 =$

EXAMPLE

We know that $(x - y)(x + 2y) = x^2 + xy - 2y^2$.
This becomes $x^2 + xy - 2y^2 = (x - y)(x + 2y)$.
The factors of $x^2 + xy - 2y^2$ are $(x - y)$ and $(x + 2y)$.

EXAMPLE

Let us now find the factors of $p^2 - 2pq - 15q^2$.

$$p^2 - 2pq - 15q^2 = (p + N_1 q)(p + N_2 q)$$
$$= p^2 + (N_1 + N_2)pq + (N_1 \times N_2)q^2$$
$$= p^2 \quad -2pq \quad + \quad -15q^2$$
$$\qquad\qquad \downarrow \qquad\qquad\qquad \downarrow$$
$$\qquad add\ to\ -2 \quad multiply\ to\ -15$$

$-1 \times 15 = -15$ ✓	$-3 \times 5 = -15$ ✓	$3 \times (-5) = -15$ ✓
$-1 + 15 = 14$ ✗	$-3 + 5 = 2$ ✗	$3 + (-5) = -2$ ✓

So, $p^2 - 2pq - 15q^2 = (p + 3q)[p + (-5)q]$
$\qquad\qquad\qquad = (p + 3q)(p - 5q).$

The factors of $p^2 - 2pq - 15q^2$ are $(p + 3q)$ and $(p - 5q)$.

Try this

10 (a) $x^2 + x - 6 =$ (d) $a^2 - ab - 6b^2 =$ (g) $c^2 - 7cd + 12d^2 =$
     (b) $x^2 + xy - 6y^2 =$ (e) $l^2 + 2lm - 3m^2 =$ (h) $r^2 + 4r - 5 =$
     (c) $a^2 - a - 6 =$ (f) $p^2 + 6pq - 16q^2 =$ (i) $p^2 + 5pq + 6q^2 =$

EXAMPLE

$(5 - 2n)(2 - 3n) = 10 - 19n + 6n^2$

This becomes,

$10 - 19n + 6n^2 = (5 - 2n)(2 - 3n)$

The factors of $10 - 19n + 6n^2$ are $(5 - 2n)$ and $(2 - 3n)$

EXAMPLE
Let us now find the factors of $4 - 5p - 6p^2$.

$4 - 5p - 6p^2 = (M_1 + N_1 p)(M_2 + N_2 p)$
$\qquad\qquad\quad = M_1 \times M_2 + (N_1 \times M_2 + N_2 \times M_1)p + N_1 \times N_2 p^2$
$\qquad\qquad\quad = \quad\;\; 4 \qquad\qquad\qquad -5p \qquad\qquad\quad -6p^2$

Now there are four numbers to find.

$M_1 \times M_2$:	$1 \times 4 = 4$ ✓	$1 \times 4 = 4$ ✓	$1 \times 4 = 4$ ✓
$N_1 \times N_2$:	$1 \times (-6) = -6$ ✓	$2 \times (-3) = -6$ ✓	$(-2) \times 3 = -6$ ✓
$N_1 \times M_2 + N_2 \times M_1$:	$1 \times 4 + (-6) \times 1 = -2$ ✗	$2 \times 4 + (-3) \times 1 = 5$ ✗	$(-2) \times 4 + 3 \times 1 = -5$ ✓

So, $4 - 5p - 6p^2 = [1 + (-2)p](4 + 3p)$
$\qquad\qquad\qquad = (1 - 2p)(4 + 3p)$

The factors of $4 - 5p - 6p^2$ are $(1 - 2p)$ and $(4 + 3p)$.

Try this

11 Factorise

     (a) $2 + x - 6x^2$ (e) $6a^2 - 5a - 6$ (i) $8a^2 - 2ab - 3b^2$
     (b) $6 + 7a - 3a^2$ (f) $8p^2 + 10p - 7$ (j) $6 - 13y + 6y^2$
     (c) $3a^2 - 7a - 6$ (g) $6q^2 - 23q - 13$ (k) $2 + 14r + 20r^2$
     (d) $5x^2 + 17x - 12$ (h) $6x^2 + 5xy - 6y^2$ (l) $6c^2 + 7cd - 20d^2$

A SPECIAL EXAMPLE

Find the factors of $x^2 - y^2$.

Compare this with: Find the factors of $x^2 + xy - 2y^2$

$x^2 + xy - 2y^2$
$= (x + N_1 y)(x + N_2 y)$ N_1 and N_2 multiply to -2, add to 1
$= (x - y)(x + 2y)$ $(-1$ and $2)$

$x^2 - y^2$ can be written as $x^2 + 0xy - y^2$.

$x^2 + 0xy - y^2$
$= (x + N_1 y)(x + N_2 y)$ N_1 and N_2 multiply to -1, add to 0
$= (x - y)(x + y)$ $(-1$ and $1)$

$x^2 - y^2 = (x - y)(x + y)$

$x^2 - y^2$ is an expression known as **the difference of two squares**.

Here are some more examples.

EXAMPLE

$4p^2 - 9q^2 = (2p)^2 - (3q)^2 = (2p - 3q)(2p + 3q)$

EXAMPLE

$1 - y^4 = (1)^2 - (y^2)^2$
$\qquad = (1 - y^2)(1 + y^2)$
$\qquad = (1 - y)(1 + y)(1 + y^2)$

EXAMPLE

$50m^2 - 72 = 2(25m^2 - 36)$ *common factor 2*
$\qquad\quad\; = 2(5m - 6)(5m + 6)$

Try this

12 Factorise

(a) $a^2 - b^2$ (d) $16x^2 - 25y^2$ (g) $a^4 - b^4$
(b) $p^2 - 4q^2$ (e) $4p^2 - q^2$ (h) $40m^2 - 10$
(c) $9c^2 - d^2$ (f) $1 - 16n^2$ (i) $-18b^2 + 98$

Consider these two cases before you try the examples which follow.

Case 1 Find the factors of $2x^2 - 2x - 24$
$2x^2 - 2x - 24 = 2(x^2 - x - 12)$ *common factor 2*
$\qquad\qquad\quad = 2(x - 4)(x + 3)$

The factors of $2x^2 - 2x - 24$ are 2, $(x - 4)$ and $(x + 3)$.

Case 2   Write $\dfrac{p^2 - pq}{p^2 + pq - 2q^2}$ in as simple a way as possible.

$p^2 - pq = p(p - q)$
$p^2 + pq - 2q^2 = (p + 2q)(p - q)$

$$\frac{p^2 - pq}{p^2 + pq - 2q^2} = \frac{p(p - q)}{(p + 2q)(p - q)}$$

$$= \frac{p}{p + 2q}$$

Try these

Factorise completely:

13 (a) $x^2 + 2x + 1$
    (b) $p^2 - 4q^2$
    (c) $a^2 - 7a + 12$
    (d) $2p^2 + 5p - 3$
    (e) $12a^2 - 11a + 2$
    (f) $8a^2 + 16ab + 6b^2$
    (g) $4x^2 + 10xy - 6y^2$
    (h) $6p^2 - 15pq + 9q^2$
    (i) $3m^2 - 27$
    (j) $6x^2 - 9x - 6$
    (k) $7 - 28m^2$
    (l) $27p^2 + 36pq + 12q^2$

14 Simply the following:

    (a) $\dfrac{12pq + 3q^2}{3q}$

    (b) $\dfrac{2ab}{2ab + 6ab}$

    (c) $\dfrac{x^2 - y^2}{x - y}$

    (d) $\dfrac{a^2 - 7a + 12}{3a - 12}$

    (e) $\dfrac{a^2 - a - 6}{a^2 - 9}$

    (f) $\dfrac{2p^2 + 5pq - 12q^2}{2p^2 - 11pq + 12q^2}$

Now try this...

Let us look at the value of $x^2 - 2x - 3$ as the value of x changes.

$x = 1$	$x^2 - 2x - 3$	$= 1^2 - 2 \times 1 - 3$	$= -4$
$x = 2$	$x^2 - 2x - 3$	$= 2^2 - 2 \times 2 - 3$	$= -3$
$x = 3$	$x^2 - 2x - 3$	$= 3^2 - 2 \times 3 - 3$	$= 0$
$x = 4$	$x^2 - 2x - 3$	$= 4^2 - 2 \times 4 - 3$	$= 5$
$x = 0$	$x^2 - 2x - 3$	$= 0^2 - 2 \times 0 - 3$	$= -3$
$x = -1$	$x^2 - 2x - 3$	$= (-1)^2 - 2 \times (-1) - 3$	$= 0$
$x = -2$	$x^2 - 2x - 3$	$= (-2)^2 - 2 \times (-2) - 3$	$= 5$

Do you think there is a minimum value of $x^2 - 2x - 3$?
Let us put the values in a table.

x	...	4	3	2	1	0	−1	−2	...
$x^2 - 2x - 3$	...	5	0	−3	−4	−3	0	5	...

$x = 3$ is a **zero** of $x^2 - 2x - 3$

$x = 1$ is the **minimum** of $x^2 - 2x - 3$.
(Are you certain -4 is the minimum value?)

$x = -1$ is a **zero** of $x^2 - 2x - 3$.

$x = 1$ is the minimum of $x^2 - 2x - 3$ for the values of x which we have tried.
You might want to use your calculator to try other values of x round about $x = 1$ ($x = 0{\cdot}5$, $x = 1{\cdot}7$ etc).

You might also want to draw a graph.

In a later unit, we explain how to find the minimum value of an expression without going through the above process. Perhaps you might think about this now.

Would factorising $x^2 - 2x - 3$ help to find zeros and minimum?

Investigate another expression such as $x^2 + x - 2$.

What is different about the expression $2 - x - x^2$?
It does not have a minimum value.
What does it have?

KEY QUESTIONS

1

K1 Factorise $4p^2q^2 - 10p^3q$.

K2 Factorise $x^2 + 7x + 10$.

K3 Factorise in full $8 - 10y - 12y^2$.

K4 Factorise $4m^2 - n^4$.

K5 Factorise in full $2 - 32a^4$.

K6 Simplify, as far as possible, $\dfrac{x^2 - 2x}{x^2 - 4}$.

EQUATION OF A STRAIGHT LINE AND INEQUALITIES

● Belinda Robertson turned a hobby into a business now worth £250,000 a year, writes **SUE PARRISH**, by proving that where there's a wool there's a way. Ruth Magee modelled Belinda's high class cashmere clothes for photographer **ERIC THORBURN**

● Karen has just completed a two year College course in textiles and knitting. She would like to be as successful as Belinda Robertson.

Karen's speciality is in making individually designed ski jerseys and snoods (snow-hoods).

It takes her three hours to make a jersey and half an hour for a snood.

It takes a lot more wool to make a jersey.

She can sell a jersey for £45 and a snood for £10.

How many jerseys and snoods should she produce each week?

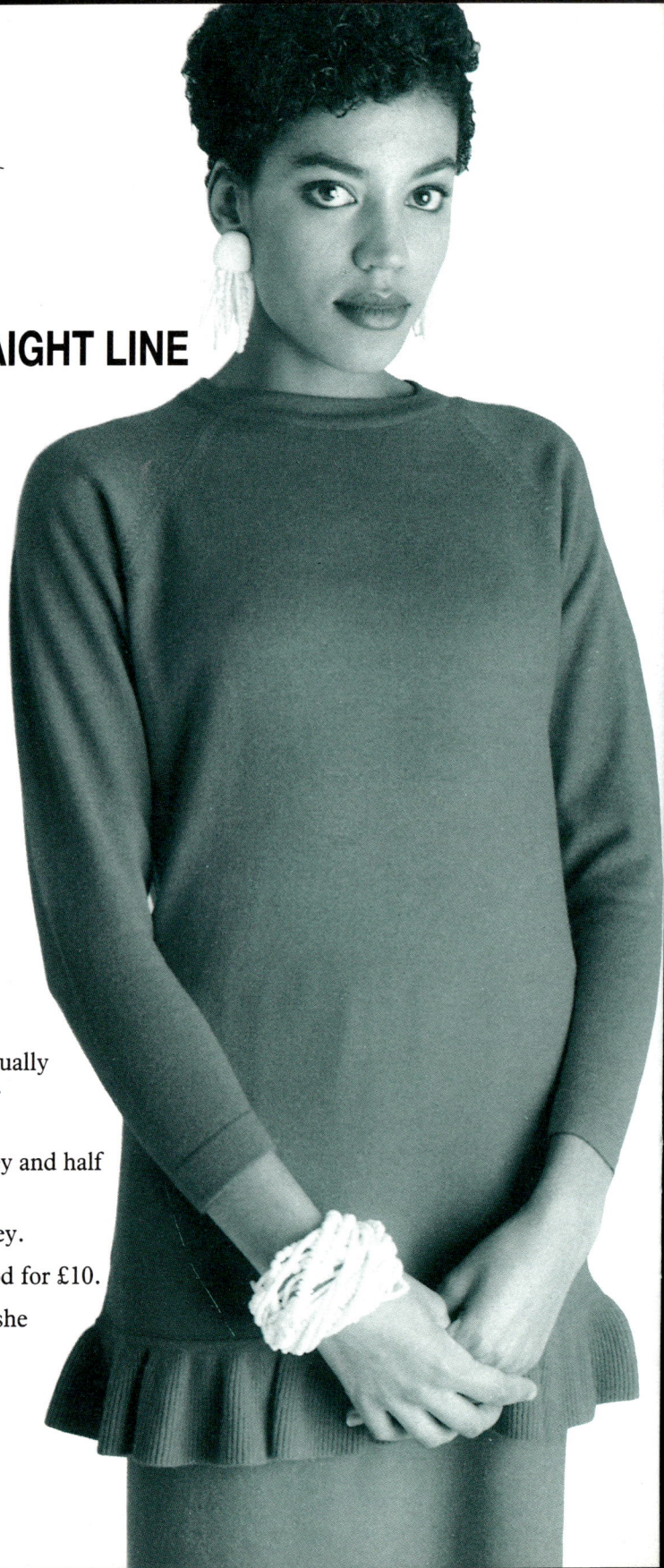

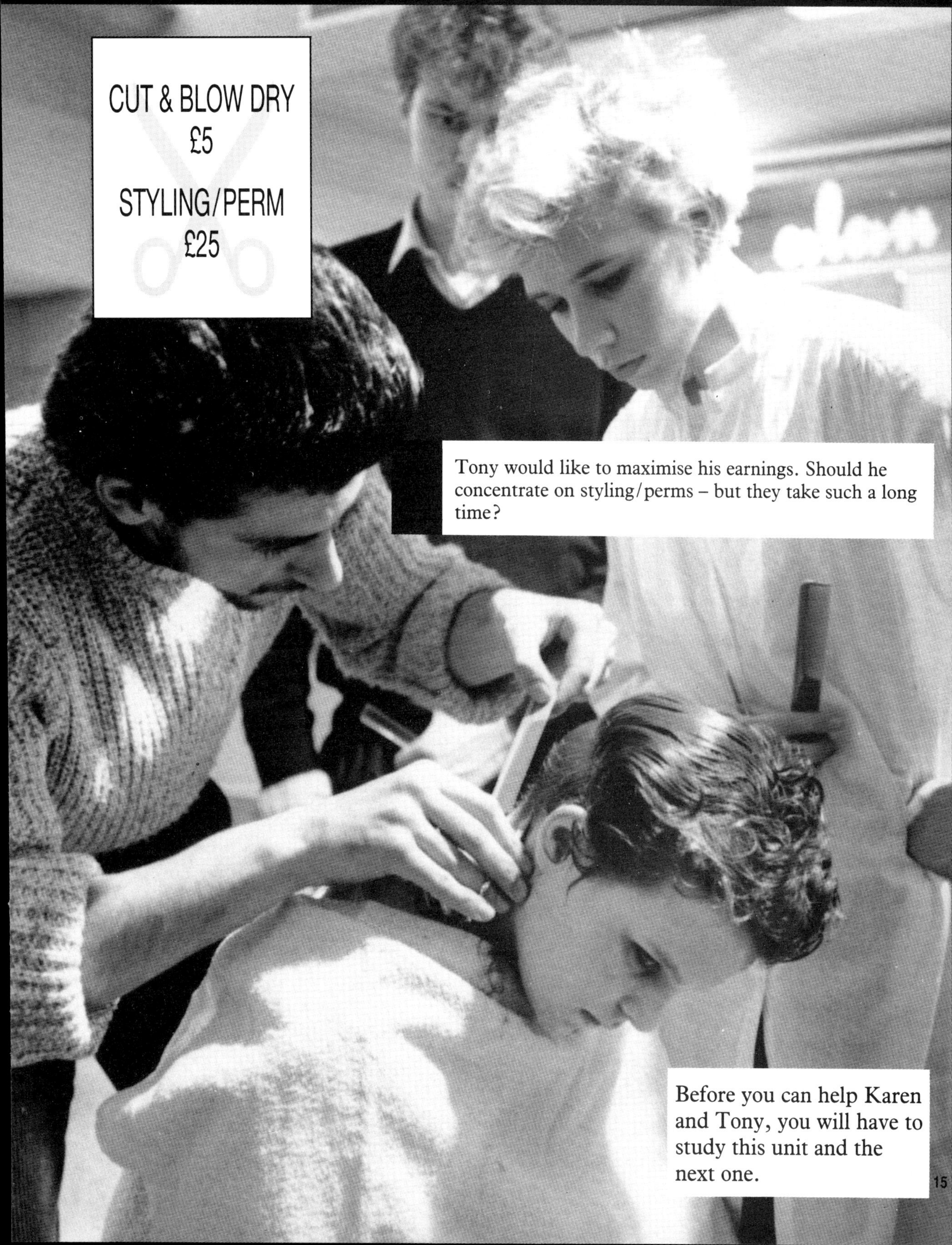

CUT & BLOW DRY
£5

STYLING/PERM
£25

Tony would like to maximise his earnings. Should he concentrate on styling/perms – but they take such a long time?

Before you can help Karen and Tony, you will have to study this unit and the next one.

15

Equation of a straight line

The first step in drawing a graph of an equation such as $2x+y=8$ is to make a table of values.

For example

x	-4	-2	0	2	4
y					

Now find the values for y by substitution into the equation $2x+y=8$.

For example, when $x=-4$

$$2x+y=8$$
$$2(-4)+y=8$$
$$-8+y=8$$
$$y=8+8$$
$$y=16$$

Then complete the table:

x	-4	-2	0	2	4
y	16	12	8	4	0

and draw the graph on a cartesian diagram.

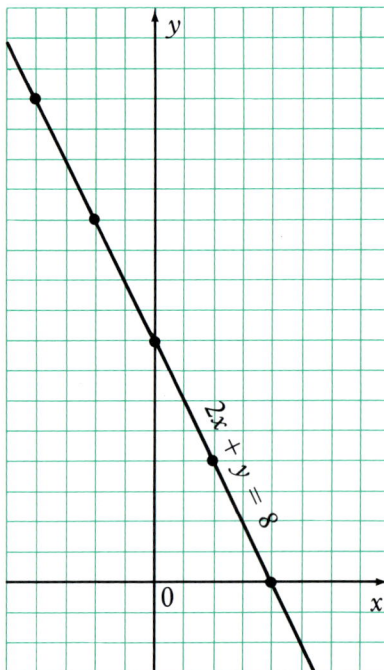

Try these

1 Draw graphs of the following equations on a cartesian diagram.
   - (a) $3x+2y=6$
   - (b) $2y=12-4x$
   - (c) $2x-y=4$
   - (d) $3x=9-y$

2 (a) In what way are the graphs you have drawn in question 1 similar?

   (b) What can you say about the graph of the equation $ax+by=c$, where a, b and c are constants (numbers)?

Equations of the form $ax+by=c$ are called **linear equations**.

The graph of the equation $ax+by=c$, where a, b and c are constants, is always a straight line. To draw a straight line, you need only two points.

Going back to the equation at the start of the unit, $2x+y=8$, the easiest points to find are when $x=0$ and when $y=0$.

x	0
y	0

$\longrightarrow$

x	0	4
y	8	0

Plot the two points (0,8) and (4,0) and join them to give the straight line. It is always advisable to check with another point, for example (2,4).

Try these

3 Draw graphs of the following equations:

    (a) $y = 3x - 6$

    (b) $2y + 3x = 12$

    (c) $2y - 3x = 6$

4 Draw graphs of the following equations:

    (a) $y = 6$ (b) $3x = 9$ (c) $y + 4 = 0$

5 Describe the graphs you have drawn in question 4.

6 Draw the following pairs of equations on the same cartesian diagram.

    (a) $x + y = 6$ and $y = 6 - x$

    (b) $y = 2x - 4$ and $2x - y = 4$

    (c) $2y - 4x = 8$ and $y = 2x + 4$

    (d) $2x + y = 4$ and $3y = 12 - 6x$

7 Describe why each pair of equations in question 6 has the same graph.

8 Match the equation in list A to the equation in list B that has the same graph.

List A	List B
$x + y = 4$	$y + 2x = 7$
$2x - y = 7$	$y = 4 - x$
$2y = 14 - 4x$	$y - x = 2$
$2y = 2x + 4$	$y = 2x - 7$

$2x - y = 6$ can be changed into the form $y = 2x - 6$. Check that they have the same graph to prove that this is true.

The steps involved in changing $2x - y = 6$ into $y = 2x - 6$ are as follows:

$$2x - y = 6$$

Step 1 (subtract $2x$) $-y = -2x + 6$

Step 2 (multiply by -1) $(-1)(-y) = (-1)(-2x) + (-1)(6)$

$$y = 2x - 6$$

Try this

9 Change the following equations into the form $y = ax + b$ where a and b are constants.

    (a) $2x + y = 6$ (b) $y - 3x = 6$ (c) $4x + 2y = 8$

    (d) $x = y + 7$ (e) $3x = 8 - 2y$ (f) $2x - 3 = y$

You will remember from unit 5 of the *Yellow Book* about **gradient**.

$$\text{Gradient} = \frac{\text{vertical distance}}{\text{horizontal distance}}$$

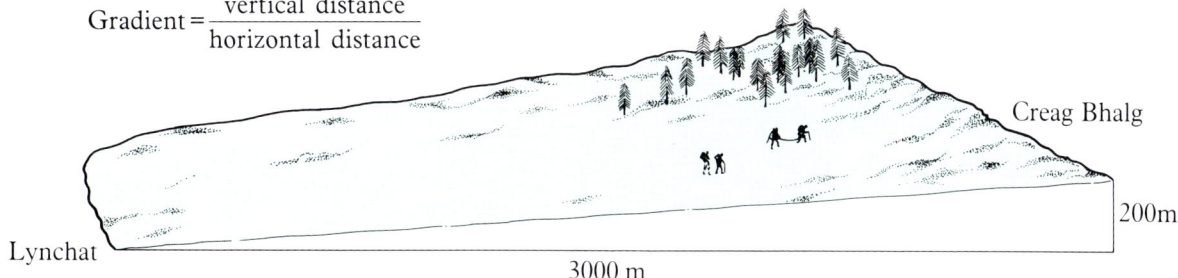

Creag Bhalg

200m

Lynchat

3000 m

The **gradient** of a straight line on a cartesian diagram is calculated in the same way.

Let us look at the line with equation $y = 2x + 3$.

$$\text{Gradient} = \frac{\text{vertical distance}}{\text{horizontal distance}}$$

$$= \frac{4}{2}$$

$$= 2$$

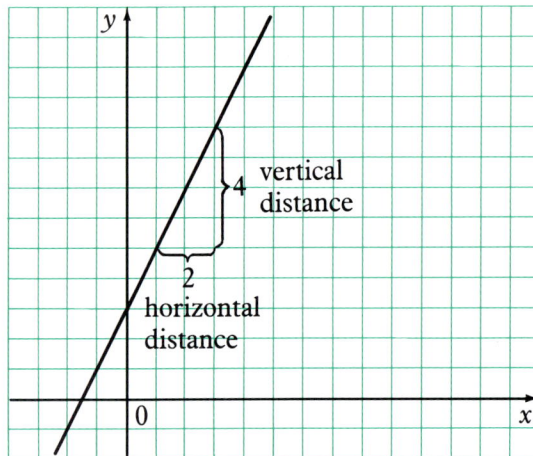

Note that the straight line with equation $y = 2x + 3$ cuts the y-axis at the point (0,3).

The number 3 is called the **intercept on the y-axis**.

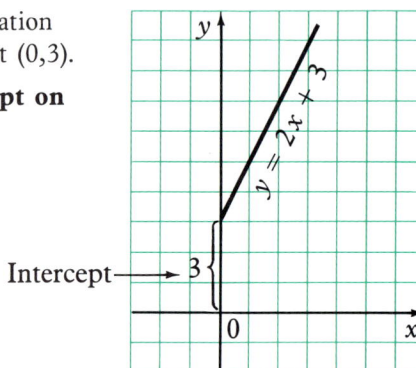

Intercept → 3

Try this

10 Calculate the gradient and intercept on the y-axis of the following lines.

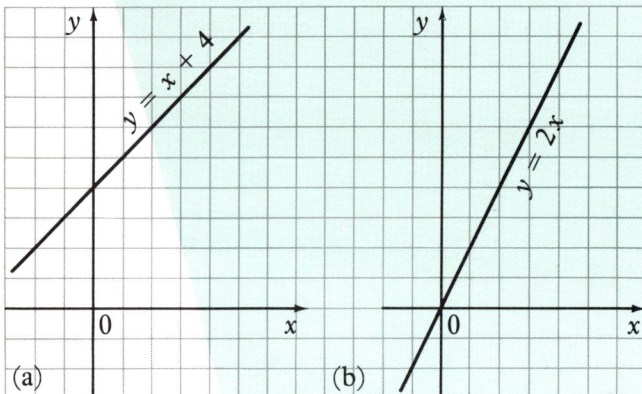

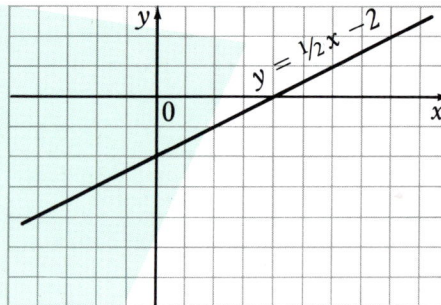

(a)   (b)

When you are out cycling, you will be in no doubt when a hill is rising or falling.

2

Gradient is positive

Gradient is negative

The convention in mathematics is that when a line **rises** from left to right it has **a positive gradient** (going up hill).

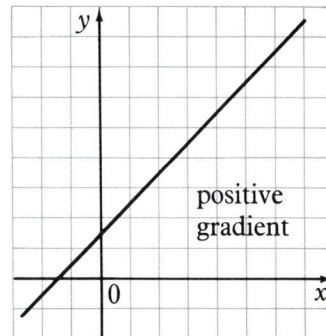

positive gradient

When a line is **falling** from left to right it has **a negative gradient** (down hill).

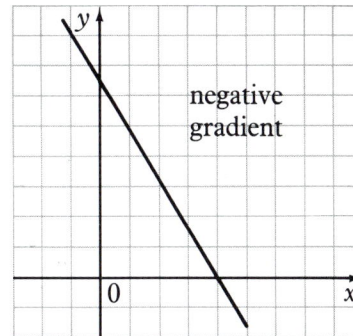

negative gradient

EXAMPLE

$$\text{Gradient} = \frac{\text{vertical distance}}{\text{horizontal distance}}$$

$$= \frac{-6}{2}$$

$$= -3$$

Intercept on y-axis is 6.

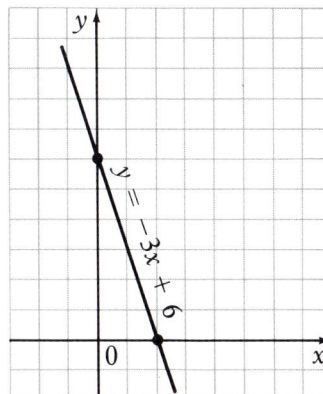

$y = -3x + 6$

19

11 State the gradients and intercept on the *y*-axis of the following lines.

(a)

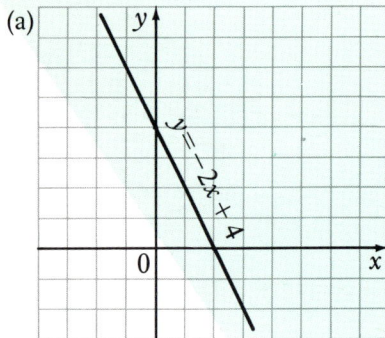

(b)

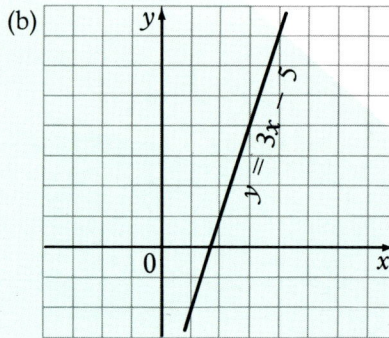

(c)

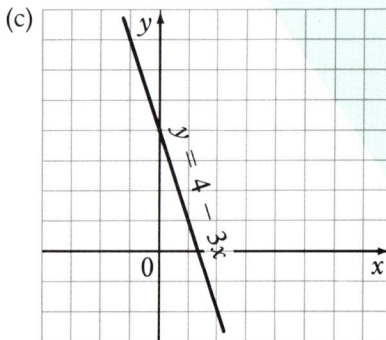

(d)

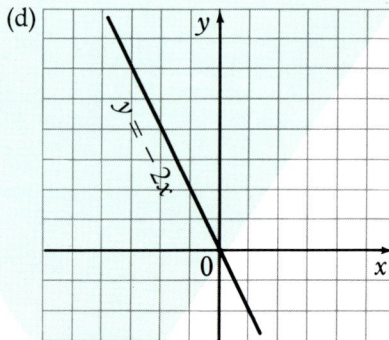

12 A line has an equation $y = ax + b$. State the gradient and intercept on the *y*-axis.

When the equation of a line is written in the form $y = ax + b$, where *a* and *b* are constants, it is possible to state the gradient and intercept on the *y*-axis without drawing a graph. In question 11(a), the equation is $y = -2x + 4$: the gradient is -2 and the intercept on the *y*-axis is 4.

$$y \; = \; \underset{\text{gradient}}{-2x} \; \underset{\substack{\text{intercept on} \\ \text{y-axis}}}{+4}$$

13 Without drawing, state the gradient and intercept on the *y*-axis of the lines whose equations are:

(a) $y = 3x - 6$ (b) $y = -\frac{1}{2}x + 4$

(c) $y = 6 - 3x$ (d) $2y = x + 6$

2

Earlier in this unit, we noticed that the equation of a line can be given in various forms.

For example,
$$\left.\begin{array}{l} 2x+y=6 \\ y=6-2x \\ 6x+3y=18 \end{array}\right\} \begin{array}{l}\text{these are all equations of the} \\ \text{same line}\end{array}$$

It is possible to state the gradient and intercept of a line so long as the equation given is changed into the form $y=ax+b$.

For example, $\qquad 6x+3y=18$
Step 1 (subtract $6x$) $\qquad 3y=-6x+18$
Step 2 (divide by 3) $\qquad y=\underset{\text{gradient is }-2}{\underline{-2x}}+\underset{\text{intercept is 6}}{\underline{6}}$

gradient is -2 $\qquad$ intercept is 6

Try these

14 Without drawing, state the gradient and intercept on the y-axis of the lines whose equations are:

(a) $x+y=-4$ $\qquad$ (b) $x=2y+7$
(c) $3x+2y=-4$ $\qquad$ (d) $3x-y=-7$

15 State the equation of the line, given the gradient and intercept on the y-axis:

(a) gradient is 2 and intercept is 4.
(b) gradient is -3 and intercept is -4.
(c) gradient is $\frac{1}{2}$ and intercept is 0.

16 For each of the three equations found in question 15, state two points which lie on the line. Do not include the intercept on the y-axis.

If you can calculate the gradient of a line and its intercept on the y-axis, you can then state its equation. All you need to know are the co-ordinates of any two points on a line and you can then work out the equation of a line.

For example,
$(-2,-1)$ and $(1,5)$ are two points on a line.

Intercept on y-axis is 3.

$$\text{gradient}=\frac{\text{vertical distance}}{\text{horizontal distance}}$$

$$=\frac{6}{3}$$

$$=2$$

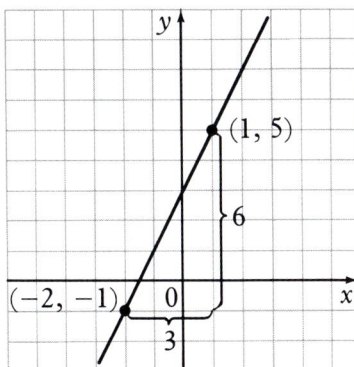

Equation of the line is $y=2x+3$.

17 Find the equations of the lines on which the following pairs of points lie.

   (a) (1,3) and (3,1)
   (b) (1,2) and (3,−2)
   (c) (4,−1) and (−4,3)
   (d) (2,−1) and (−1,−10)

18 State the equations for question 17 in a different form.

19 (a) Draw graphs on the same diagram of the following pairs of equations:
     (i) $y = 2x + 4$ and $y = 2x - 2$
     (ii) $y = \frac{1}{2}x + 5$ and $y = \frac{1}{2}x - 3$

   (b) What can you say about lines which have the same gradient?

Inequations

A line drawn on a cartesian diagram divides it into three regions.

Let us investigate the regions on either side of a line.

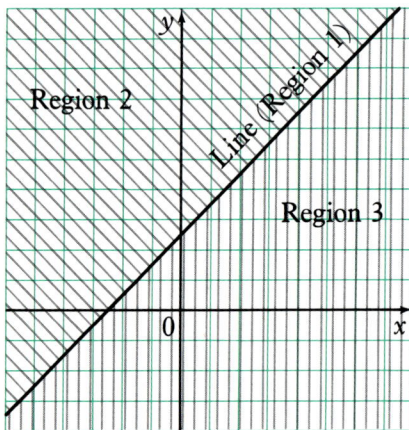

The line $2x + y = 8$ divides a cartesian diagram into three regions:

   (i) the set of points which make up the line ($2x + y = 8$);
   (ii) region A; and
   (iii) region B.

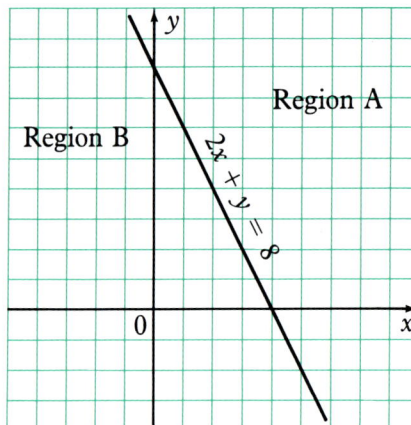

Let us examine region B. You could shade in region B by drawing a lot of lines that are parallel to $2x+y=8$.

A few lines are shown as examples.

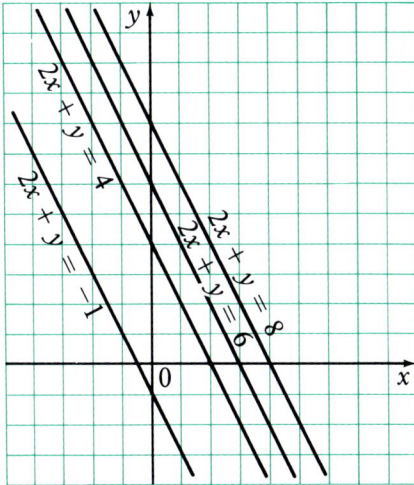

Region B could be described as:

$2x+y$ **is less than** 8
(i.e. $2x+y<8$).

Select a few points in region B to confirm that, for the region, $2x+y$ is less than 8 (i.e. $2x+y<8$)

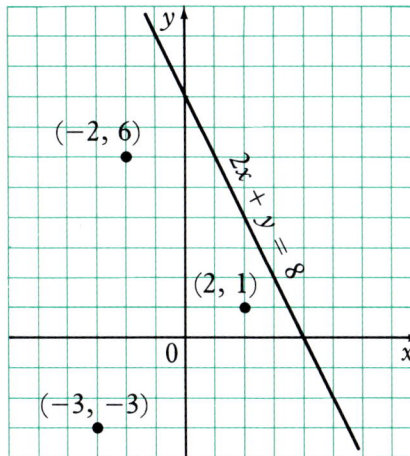

For $(2,1)$: $2x+y=5$ <8

For $(-2,6)$: $2x+y=2$ <8

For $(-3,-3)$: $2x+y=-9$ <8

It would now be fair to claim that for all the points in region B:

$$2x+y<8$$

Likewise, region A could be described as:

$2x + y$ is greater than 8

(i.e. $2x + y > 8$)

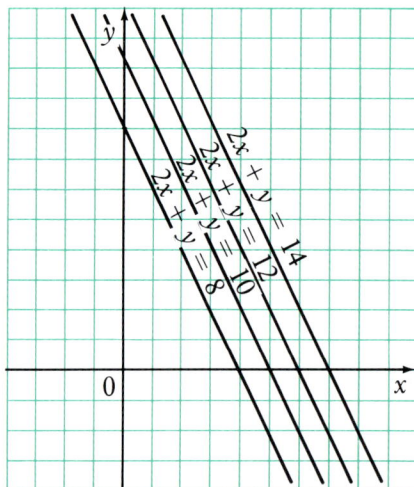

Again you could confirm this by selecting a few points.

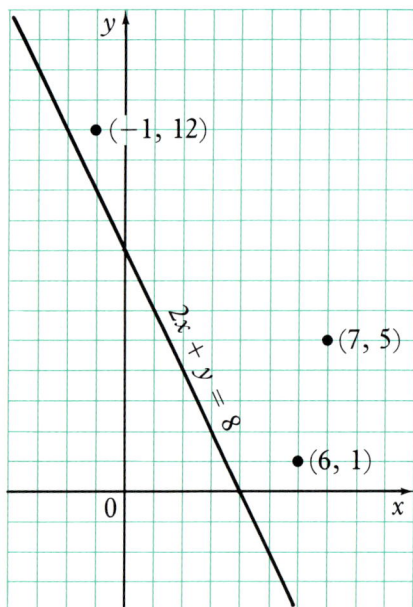

For (6,1): $2x + y = 13$ > 8

For (7,5): $2x + y = 19$ > 8

For $(-1, 12)$: $2x + y = 10$ > 8

So, for all the points in region A,

$$2x + y > 8$$

2

The line $2x + y = 8$ divides a cartesian diagram into 3 regions:

(i) line $2x + y = 8$
(ii) $2x + y > 8$
(iii) $2x + y < 8$

All lines divide a cartesian diagram in the same way.

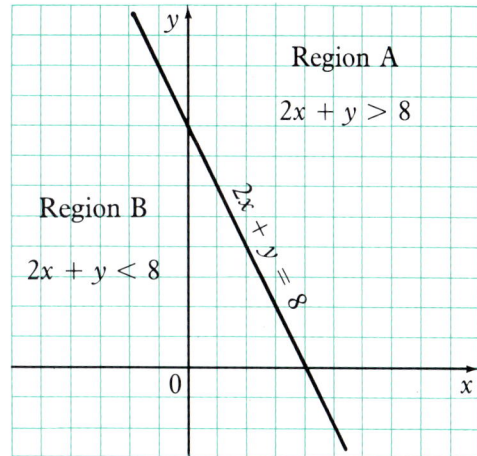

Try this

20 Draw graphs of the following equations on separate cartesian diagrams. By using $<$ or $>$, describe the regions on either side of the lines you have drawn. (Remember to check by selecting a number of points.)

(a) $2y + x = 6$ (b) $y = 2x - 3$ (c) $3x - 2y = 12$

We are familiar with the symbols $=$, $<$ and $>$. In the future you may meet a combination of these symbols such as $\leqslant$ or $\geqslant$.

$2x + y \leqslant 8$ means that $2x + y$ is **less than or equal to** 8. That is, the solution set of the statement $2x + y \leqslant 8$ is all of the points (x, y) where $2x + y$ is less than or equal to 8.

Likewise $\geqslant$ means **greater than or equal to**.

The diagram shows the solution set for $2x + y \leqslant 8$. (The line is included in the solution set.)

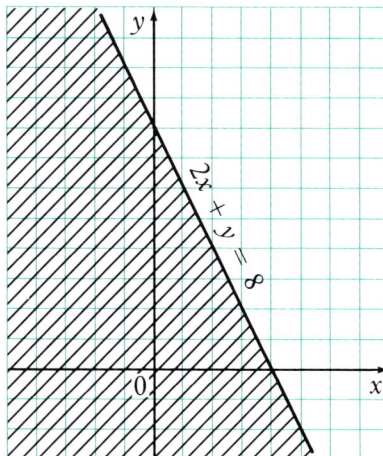

Note the use of the phrase 'solution set'. An inequation such as $2x + y \leqslant 8$ does not have just one solution.

Points such as $(1,1)$, $(2, -5)$, $(-4, -7)$, etc are all part of the solution set of $2x + y \leqslant 8$.

This diagram shows the solution set for $2x + y > 8$.

Note the use of a broken line because the solution set does not include the points where $2x + y = 8$.

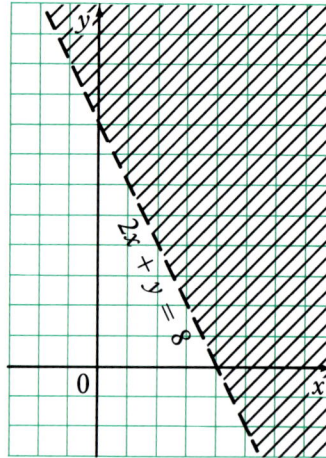

Try this

21 Show the solution sets for the following statements on separate cartesian diagrams:

(a) $y \leqslant \frac{1}{2}x - 2$ (b) $2y - x \geqslant 6$ (c) $2y < 12 - 3x$

You will remember from earlier in the unit that the equation of a line can be given in many different forms.

The equations
$$2x + y = 8$$
$$y = 8 - 2x$$
$$16 - 2y = 4x$$
$$8 - 2x - y = 0$$
all represent the same line.

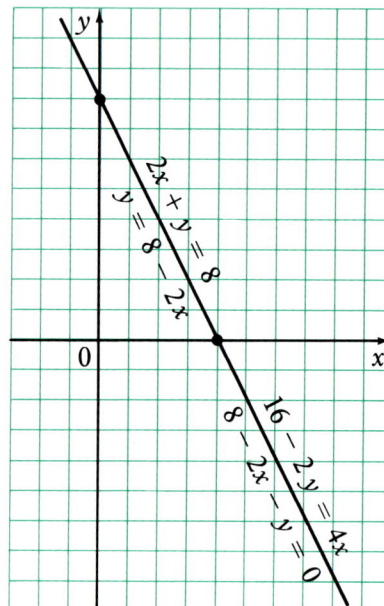

2

The regions on either side of the line could be described as shown on the diagram.

Note that this means it is very important to select a few points to check whether you should use $<$ or $>$.

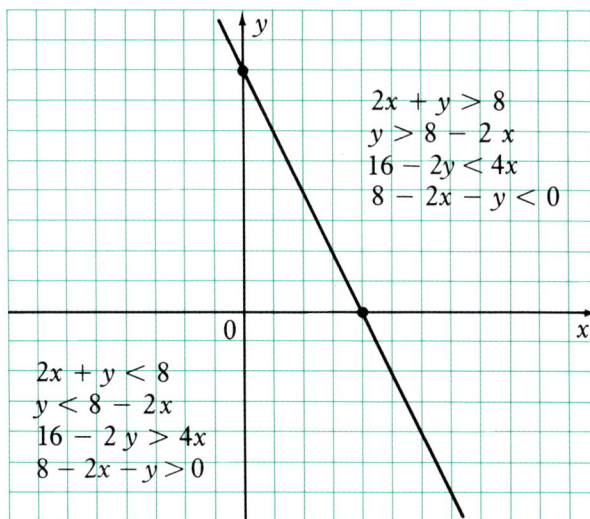

$2x + y > 8$
$y > 8 - 2x$
$16 - 2y < 4x$
$8 - 2x - y < 0$

$2x + y < 8$
$y < 8 - 2x$
$16 - 2y > 4x$
$8 - 2x - y > 0$

Try this

22 On separate cartesian diagrams show the solution sets of the following inequalities.

(a) $3x - 2y < -6$ (b) $4 - x \geqslant 2y$ (c) $x - 2y + 8 < 0$

The equation of the line on the diagram opposite could be stated in various forms such as

$$x = -4$$
$$-x = 4$$

By selecting points such as $(-6, 2)$ and $(2, -2)$, you can decide how to describe the regions on either side of the line.

For $(-6, 2)$, $-6 < -4$, so $x < -4$
But for $(-6, 2)$, $-(-6)$? 4
                       $6 > 4$
     so $-x > 4$

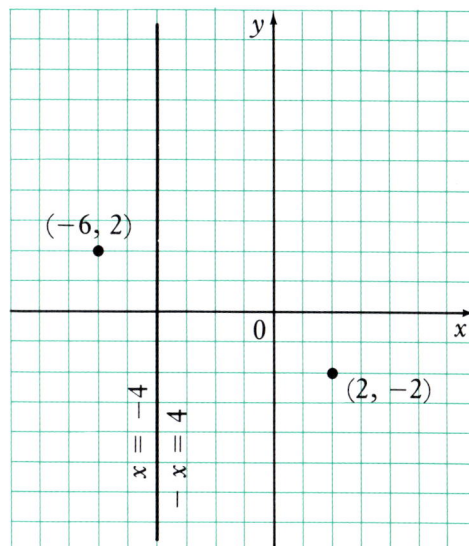

$(-6, 2)$

$(2, -2)$

This is a rather interesting result:

$x < -4$ and $-x > 4$ represent the same solution set.

Likewise, $x > -4$ and $-x < 4$ represent the same solution set.

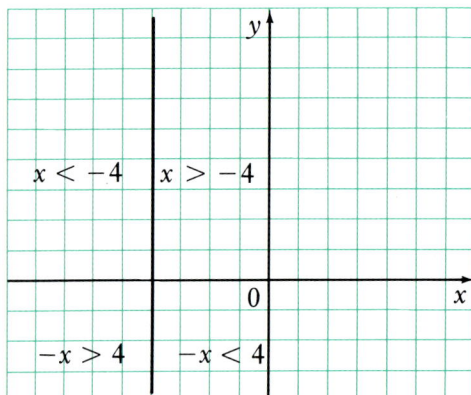

You might consult your teacher if you do not follow this.

We shall make use of this result later in the unit.

Try this

23 Draw the following pairs of equations on a separate cartesian diagram for each pair. Describe the regions which lie on either side of each line.

(a) $-3y = 6$
    $y = -2$

(b) $-4x = -4$
    $x = 1$

(c) $3x = -9$
    $x = -3$

Solving inequalities

Solving inequalities with one variable (one unknown) is exactly the same as solving equations, with one exception. The clue to that exception is contained in your answers to questions **23**(b) and **23**(c).

The answer to question 23(c) is

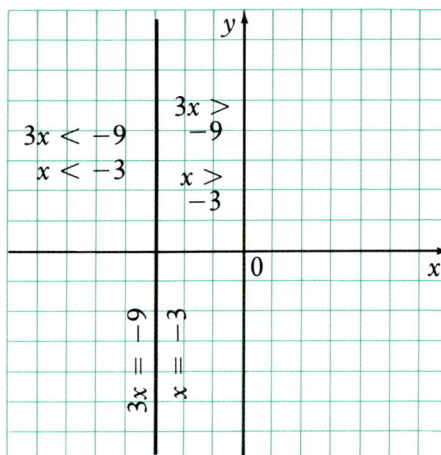

Let us examine these answers:

$3x < -9$ $3x = -9$ $3x > -9$ (divide by 3)

$x < -3$ $x = -3$ $x > -3$

It appears from the above examples that inequalities behave like equations when each side is divided by a positive number (in this case 3).

But, question 23(b)

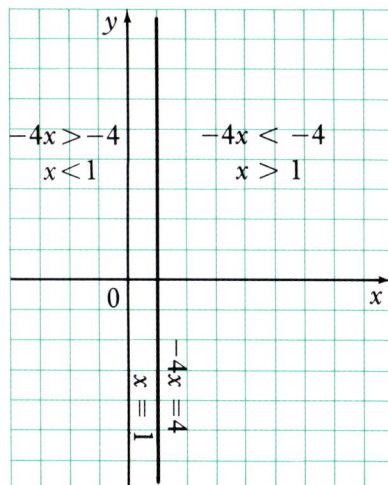

Let us examine these answers:

$-4x > -4$ $-4x = -4$ $-4x < -4$ (divide by -4)

$x < 1$ $x = 1$ $x > 1$

Notice the difference. When you divide both sides of an inequality by a negative number (in this case -4), the inequality symbol changes:

i.e. $-4x > -4$ becomes $x < 1$

        $-4x < -4$ becomes $x > 1$

EXAMPLE

Solve for x when

         $3x + 6 > x + 2$

Step 1 $3x - x > 2 - 6$

Step 2 $2x > -4$

Step 3 $x > -2$

EXAMPLE

Solve for x when

         $2x - 4 < 3x - 6$

Step 1 $2x - 3x < -6 + 4$

Step 2 $-(1)x < -2$

Remember the exception. Here divide both sides by -1.

Step 3 $x > 2$

Note the change of symbol.

EXAMPLE

$$2(3+x)-3(2x-2)\leqslant 0$$
$$6+2x-6x+6\leqslant 0$$
$$-4x+12\leqslant 0$$
$$-4x\leqslant -12$$
$$x\geqslant 3$$

Note the change of symbol.

Try this

24 Solve for x

(a) $3x<6$

(b) $-4x>-20$

(c) $2x-6\leqslant 0$

(d) $8-2x>-4$

(e) $4-2x<8-3x$

(f) $2(3x-2)>8$

(g) $4x-3>3(x-6)$

(h) $2(2x+3)+3(x+1)\leqslant -5$

More on solution sets

Linear equations of the form $ax+b=0$ have a unique solution.

Inequalities of the form $ax+b\leqslant 0$ have numerous possible solutions. The collection of all possible solutions is called the **solution set**.

When solving an equation such as $8x=19$, the main concern is about the degree of accuracy required,

e.g. $8x=19$
$x=2\cdot 375$

Possible answers are 2, 2·4, 2·38, or 2·375, depending on the required degree of accuracy. In this case, if an answer to 1 decimal place is required, the answer would be 2·4.

In a similar way, our answers to inequalities so far in this unit are not complete. To avoid complications, care was taken to ensure that each solution involved integers only. In stating the solution set of an inequality, it is of value to indicate which number system is involved.

The principal number systems are as follow.

The integers (Z) These are positive and negative whole numbers and zero,

i.e. $\{\ldots, -3, -2, -1, 0, 1, 2, 3, \ldots\}$

Rational numbers (Q) These are numbers that can be expressed as a fraction with an integer numerator and an integer denominator. These include numbers such as $3\frac{1}{4}$, $-\frac{199}{43}$, $2\cdot 1\left(\frac{21}{10}\right)$, etc.

Real Numbers (R)   These numbers include integers, rational numbers and irrational numbers. Irrational numbers are numbers such as $\sqrt{11}$, which you will meet in unit 4 in this book.

When solving $8x < 19$, the solution set will depend on the number system involved.

When using either the rational number or the real number system, the solution could be stated as being $x < 2 \cdot 375$.

However, in the integer number system, the solution could be given as being $x \leqslant 2$, where x is an integer.

As always in mathematics, it is helpful to give such solutions in a shorthand form.

For example, solve $8x < 19$ where x is an integer.

Solution set is $\{x : x \leqslant 2, \ x \in Z\}$

This is translated as:

The set of numbers x such that x is less than or equal to 2 and x is an integer.

$\in$ is shorthand for 'is a member of'.
$\{\,\}$ is shorthand for solution set.

Stating solution sets in this shorthand way is called 'using set notation'.

Try this

25   Solve the following inequalities. State the solution sets using set notation, with reference to the number systems indicated.

(a) $5x < -8, \ x \in Z$

(b) $-5x > 8, \ x \in R$

(c) $3x - 2 < 8, \ x \in Z$

(d) $8x - 3 > -16, \ x \in R$

(e) $3(2x - 1) + 4(2 - x) < 2, \ x \in Z$

(f) $5(3 - x) - 2(2x + 4) > -7, \ x \in Z$

KEY QUESTIONS

K1   A line has equation $2x - 3y + 6 = 0$. State the gradient and intercept on the y-axis of this line.

K2   A line passes through the points $(-1, 4)$ and $(1, 0)$. State the equation of this line.

K3   On a cartesian diagram, show where $y > 2 - x$.

K4   Solve for x
$$5 - 4x \leqslant 17$$

Now try this...

A   These three lines are parallel

$2x - y = 3$

$2x - y = 0$

$2x - y = -3$

They have the same gradient which is 2.
Check that this is true.

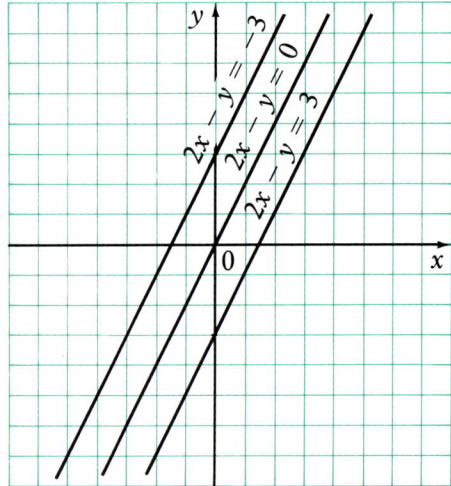

These two lines

$2x - y = 3$

$x + 2y = 4$

are at right angles to each other
(perpendicular).

Calculate the gradients of the two lines. Is
there a relationship between the gradients?

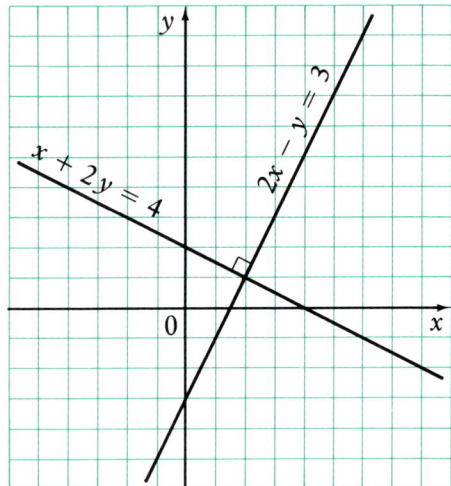

Can you prove if this relationship is true for all lines that are
perpendicular? Start with lines that pass through the origin.

B   Starting with a rod 1 m long, describe where two cuts could
be made so that the three pieces produced can form a
triangle.

SIMULTANEOUS EQUATIONS AND LINEAR PROGRAMMING

In a unit in the *Red Book* we met pairs of equations like these:

$$2x + 3y = 13$$
$$3x - y = 3$$

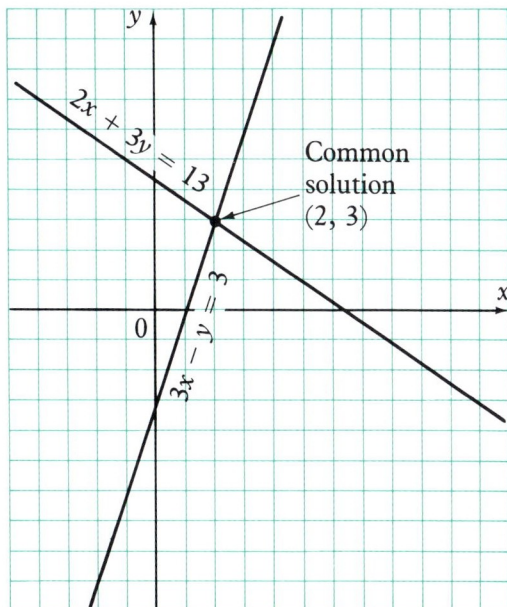

They are normally called **simultaneous equations**.

In the *Red Book*, we found the common solution of the pairs of equations by drawing graphs.

The common solution is the point of intersection [in this case (2,3)] of the two lines.

So, $x = 2$
$y = 3$

Try this

1 By drawing graphs find the common solution of these pairs of simultaneous equations.

(a) $x + y = 4$
     $2x - 3y = 3$

(b) $2x + y = -3$
    $x - 3y = 21$

This method is perfectly acceptable when the solutions are integers. When the values of x and y are not integers, you should estimate the values to 1 decimal place.

Try these

2 Solve each of these pairs of simultaneous equations graphically

(a) $x + y = 6$
     $2x - 4y = -9$

(b) $2x + 2y = -1$
    $3x - 4y = 9$

3 Why does this pair of equations not have a common solution?

$3x - y = 3$
$6x - 2y = -4$

Now that estimating solutions are involved, we shall have to find a more accurate method for establishing the common solution of a pair of simultaneous equations.

Look at these examples.

EXAMPLE

Here are two equations: (1) $x+y=5$
(2) $x-y=-1$

Watch what happens when you add equations (1) and (2).

Equations (1) + (2): $(x+y)+(x-y)=5+(-1)$

These cancel out

$$x+y+x-y=5-1$$
$$2x=4$$
$$x=2$$

Now that we know that $x=2$, we can find y from equation (1) or (2).

(1) $x+y=5$
$2+y=5$
$y=5-2$
$y=3$

So, $x=2$
$y=3$

EXAMPLE

Here are two equations: (1) $2x+3y=3$
(2) $2x-5y=11$

This time try subtracting equation (2) from equation (1).

(1)$-$(2): $(2x+3y)-(2x-5y)=3-11$

These cancel out

$$2x+3y-2x+5y=-8$$
$$8y=-8$$
$$y=-1$$

Substitute $y=-1$ in equation (1) or (2)

(1) $2x+3y=3$
$2x+3(-1)=3$
$2x-3=3$
$2x=3+3$
$2x=6$
$x=3$

So, $x=3$
$y=-1$

3

EXAMPLE

Here are two equations: (1) $4x + y = 16$
                         (2) $x + 2y = 11$

Trying to add or subtract these equations immediately will not help.

Let us try to eliminate one of the letters, say y. As we have $2y$ in the second equation, we need to have $2y$ in the first as well.

This can be done by multiplying both sides of equation (1) by 2.

$$2(4x + y) = 2 \times 16$$
$$8x + 2y = 32$$

Now subtract equation (2) from this new equation, and y is eliminated.

$$(8x + 2y) - (x + 2y) = 32 - 11$$

These cancel out

$$8x + 2y - x - 2y = 21$$
$$7x = 21$$
$$x = 3$$

Knowing that x is 3, we can find the value of y from either equation (1) or (2).

Equation (2): $x + 2y = 11$
$$3 + 2y = 11$$
$$2y = 11 - 3$$
$$2y = 8$$
$$y = 4$$

So, $x = 3$
         $y = 4$

Try this

4 Solve the following simultaneous equations.

(a) $2x + 3y = 15$
     $x - 3y = 3$

(b) $5x - 2y = 1$
     $6x - 2y = 2$

(c) $2a - 3b = 15$
     $2a + b = 3$

(d) $a - 2b = 1$
     $2a + 5b = 20$

(e) $6x + 3y = 2$
     $x + y = 1$

(f) $2a - b = 10$
     $3a + b = 12$

(g) $3s + 2t = 1$
     $2s - 3t = 18$

(h) $3a + 2b = 8$
     $7a + 5b = 20$

(i) $5x - 3y - 14 = 0$
     $3x + 2y + 3 = 0$

(j) $2a + 1 + b = 0$
     $12 + 3a - 2b = 0$

Linear programming

Let us now extend some of the work started in unit 2.

The clear region, including its boundaries, is defined by the set of inequations

$x \geqslant 0, \quad y \geqslant 0$
$2x + y \leqslant 12, \quad x + 2y \leqslant 10$

All the points in the region satisfy the four inequations.

Note that this time reverse shading has been used.

Listed below are the points within this boundary

(0,5)
(0,4) (1,4) (2,4)
(0,3) (1,3) (2,3) (3,3) (4,3)
(0,2) (1,2) (2,2) (3,2) (4,2) (5,2)
(0,1) (1,1) (2,1) (3,1) (4,1) (5,1)
(0,0) (1,0) (2,0) (3,0) (4,0) (5,0) (6,0)

The maximum value of $x + y$ within the region is given by the points (4,3) and (5,2).

So within this region, the maximum value of $x + y$ is 7.

The minimum value of $x + y$ within the region is 0.

Note: In this unit we shall use only integer values.

Another way to find the maximum or minimum value of $x + y$ in the region is to take a 'searchline'

$x + y = k$

Draw the lines for $k = 0, 1, 2, 3$ etc, until you establish which lines just pass through the region at either extremes.

The lines are $x + y = 7$ and $x + y = 0$.
So, the maximum value of $x + y$ is 7 and the minimum value of $x + y$ is 0.

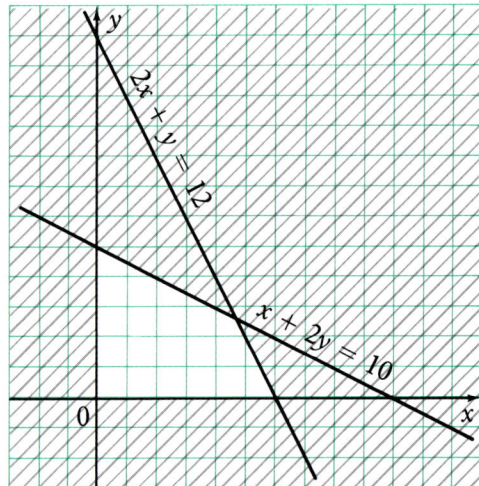

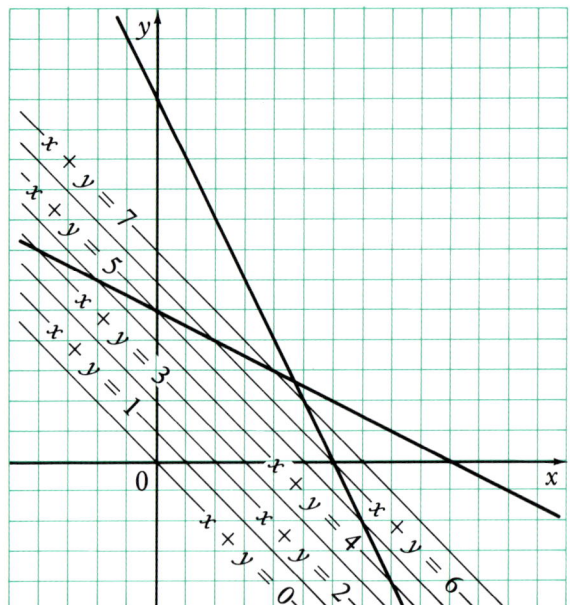

Try these

5 (a) Find the maximum and minimum values of $x+y$ in the clear region of this diagram.

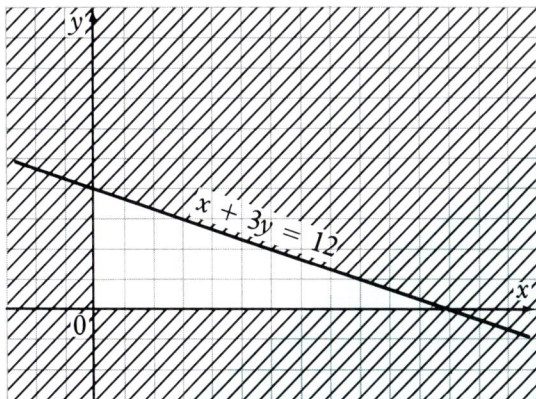

(b) Describe the clear region using inequations.

6 (a) Find the maximum and minimum values of $x+2y$ in the clear region.

Use as a searchline

$x+2y=k$ for $k=0, 1, 2$, etc.

(b) Describe the clear region.

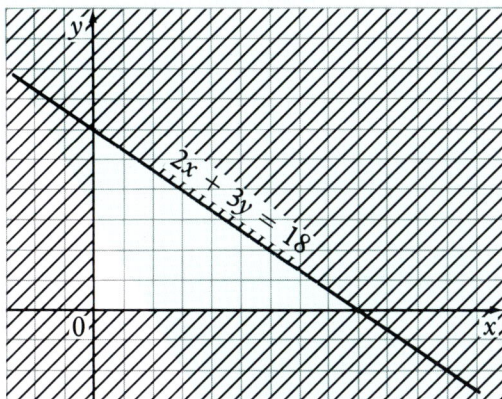

7 (a) Describe the clear region.

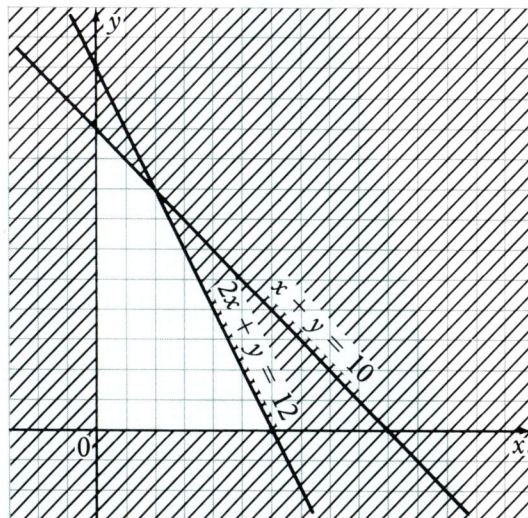

(b) Find in this region, the maximum and minimum values of $2x+y$.

8 On separate diagrams, illustrate the (clear) regions defined by the following sets of inequations.

(a) $x\geqslant0$, $y\geqslant0$, $x\leqslant6$ and $y\leqslant5$
(b) $x\geqslant0$, $y\geqslant0$, $x\leqslant5$ and $x+2y\leqslant8$
(c) $x\geqslant0$, $y\geqslant0$, $2x+y\leqslant8$ and $x+2y\leqslant8$
(d) $x\geqslant0$, $y\geqslant0$, $x+y\leqslant9$ and $2x+y\leqslant12$

9 (a) Show the region defined by:
$x\geqslant0$, $y\geqslant0$, $x\leqslant5$ and $y\leqslant6$

(b) Find in this region the maximum and minimum values of $x+2y$

10 (a) Show the region defined by:
$x\geqslant1$, $y\geqslant2$, $x+y\leqslant7$ and $2x+3y\leqslant18$

(b) Show that the maximum value of $4x+y$ is 22, and the minimum value is 6.

Linear programming — practical applications

Let us look at some applications of linear programming. At the end of this part of the unit, you might understand the meaning of the term 'linear programming'.

The hairdressing stylist has two types of customers
— cut and blow dry
— styling/perm.

A new styling or perm takes a good deal longer and costs a lot more than a 'cut and blow dry'.

On average, a cut and blow dry takes about 20 minutes for a stylist, with the assistance of a junior for 10 minutes.

A styling/perm takes two hours for a stylist, with the assistance of a junior for half an hour.

Let us put this information into a table.

	Stylist	Junior	Cost
Cut & Blow Dry	20 minutes	10 minutes	£5
Styling/Perm	2 hours	$\frac{1}{2}$ hour	£25

A stylist works 35 hours per week. A junior assists more than one stylist, and has some general cleaning duties. On average, a stylist has assistance from a junior for 10 hours per week.

What the stylist would like to know is how many cut & blow drys and how many styling/perms he should do each week, so as to maximise the amount of money earned.

Let the number of cut & blow drys per week be x.
Let the number of styling/perms per week be y.

STYLIST TIME

The stylist works 35 hours each week.
A cut & blow dry takes 20 minutes
($\frac{1}{3}$ hour) on average.
If he has x customers, then the total number of hours spent on cut & blow drys is $\frac{1}{3}x$.

A styling/perm takes 2 hours.
If he has y customers, then the total number of hours spent on styling/perms is $2y$.

So, $\frac{1}{3}x + 2y \leqslant 35$

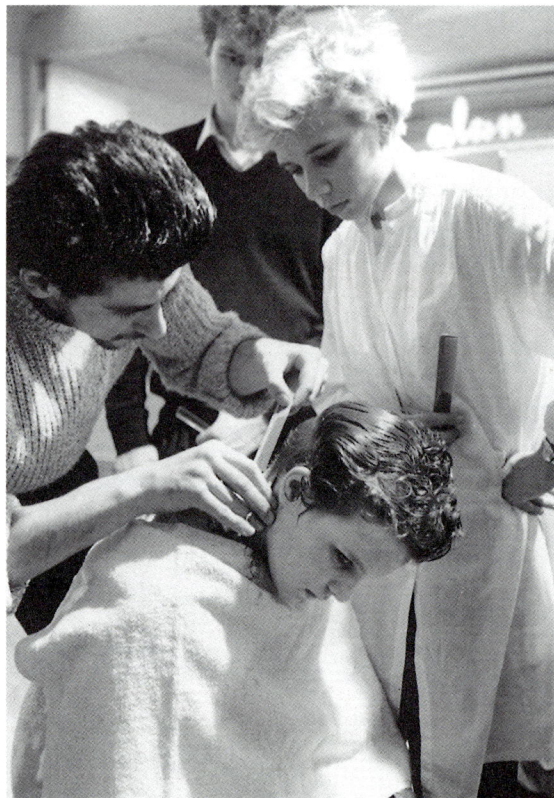

JUNIOR TIME

The junior assists a stylist for 10 hours per week.
A cut & blow dry requires 10 minutes ($\frac{1}{6}$ hour) of assistance.
Total number of hours spent on cut & blow dry is $\frac{1}{6}x$.

A styling/perm requires $\frac{1}{2}$ hour of a junior's time.
Total number of hours of assistance with styling/perms is $\frac{1}{2}y$.

So, $\frac{1}{6}x + \frac{1}{2}y \leqslant 10$

We have a set of inequations:
$\frac{1}{3}x + 2y \leqslant 35$
$\frac{1}{6}x + \frac{1}{2}y \leqslant 10$

Number of customers cannot be negative, so $x \geqslant 0$, $y \geqslant 0$

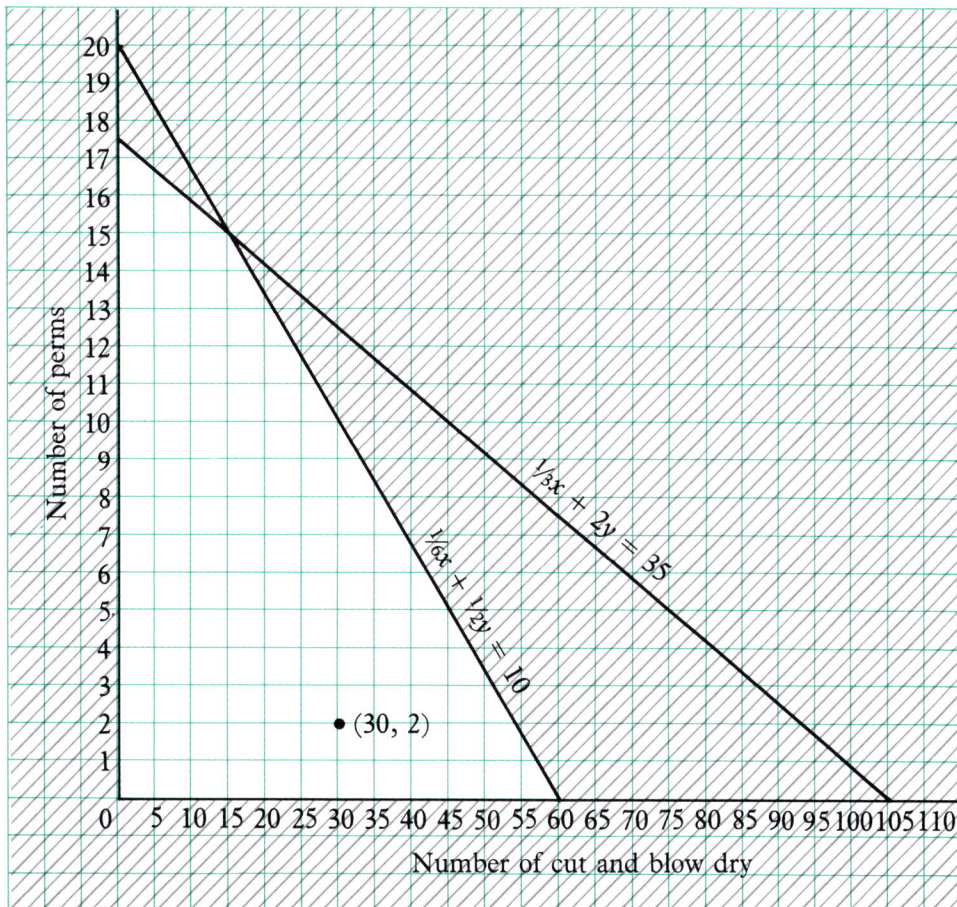

Any point within the clear region [e.g. (30, 2)] would ensure that neither the stylist nor the junior exceed the total time available.

Let us examine further the point (30, 2) which we have selected as an example.

For this point $x = 30$ and $y = 2$.
This means the number of cut & blow drys is 30 and styling/perms is 2.
This would mean that the money earned is
$30 \times £5 + 2 \times £25 = £200$

However, what we want to know is the combination of customers which gives the maximum amount of money.

Let us try a searchline. In general terms, the total amount of money taken is $5x + 25y$.

Draw lines $5x + 25y = k$, with increasing values of k, until a line just cuts into the clear region.
e.g. $5x + 25y = 375$
$5x + 25y = 400$
$5x + 25y = 425$
$5x + 25y = 450$

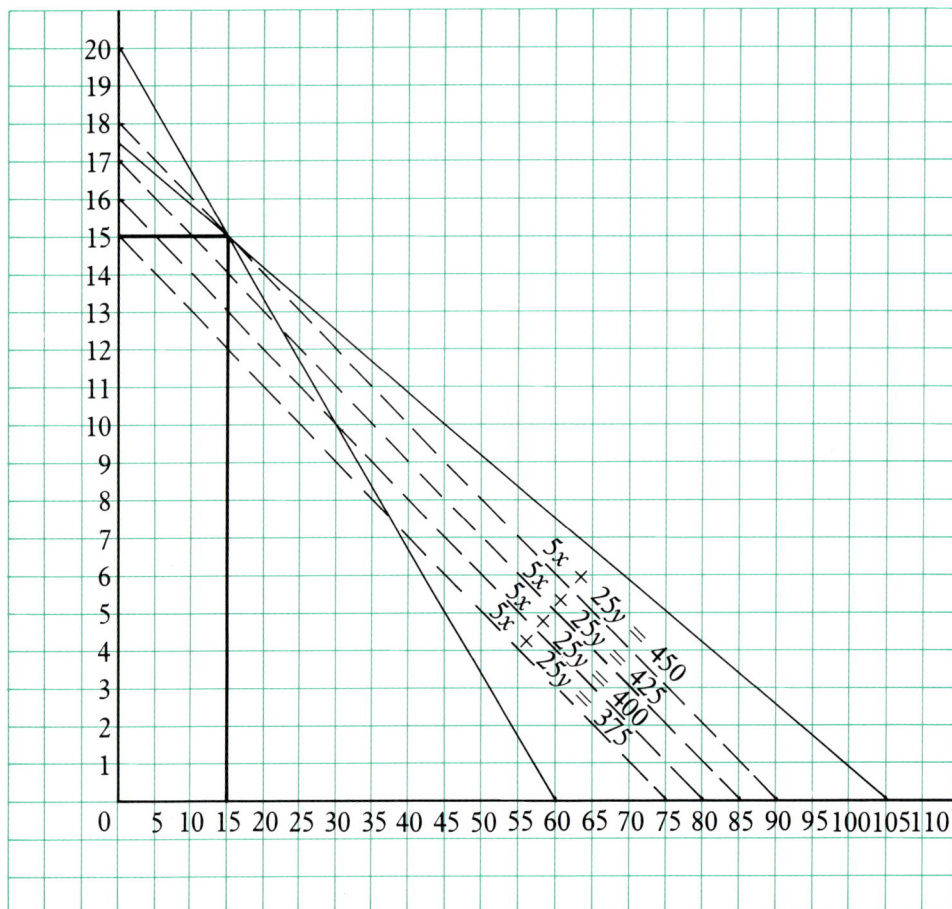

The line with equation $5x + 25y = 450$ just cuts the clear region.
Where this line cuts the clear region, $x = 15$ and $y = 15$.

3

So, number of cut & blow drys should be 15
number of styling/perms should be 15.

So the maximum possible earnings is
$15 \times £5 + 15 \times £25 = £450$

If this stylist could arrange it, he should do 15 cut & blow
drys each week, and 15 styling/perms.

Try these

11 Planning permission has been obtained for
the building of a new dog/cat home to
accommodate a maximum of 60 animals.
A dog requires 20 m² of floor space and a
cat 12 m². The total floor space available is
1000 m².
The daily boarding charges for a dog are
£3 and £2 for a cat.
How much floor space should be allocated
to dogs and how much to cats, so that
there is the potential of earning the
maximum amount from boarding charges?

(a) State inequations derived from the
information above.

(b) Show these inequations on a cartesian
diagram.

(c) How many spaces should there be for
dogs and cats so that potential earnings
are maximised.

(d) What is the maximum possible earnings
per day from boarding charges?

(e) What other factors so far not mentioned
might influence the decision on the
allocation of floor space?

12 A butcher's shop is famous for its
beefburgers and bangers.
Both the burgers and the bangers are made
from beef and animal fat, with only a trace
of other elements.
Beefburgers sell at £3·80 per kg, and
bangers at £3 per kg.

There is 170 kg of beef and 40 kg of
animal fat available to make burgers and
bangers each day.
How many kg's of each should be made to
maximise the possible earnings from sales?

WE GUARANTEE

Beefburgers
90%
pure beef

Bangers
at least 70% beef

Now try this...

Karen has just completed a two year College course in textiles. She is determined to establish her own business. Her speciality is knitting. A major ski-ware distributor is prepared to buy her individual designed ski-jerseys for £45 and snoods (snow hoods) for £10 each.

Using her knitting machine, she can produce a jersey in 3 hours, and a snood in $\frac{1}{2}$ hour

She is going to work from home. Accommodation is a bit of a problem. She only has room for 15 boxes of wool at a time. Completed items take up the same space as the balls of wool. Each box contains 20 balls of wool. A ball of wool costs £1.

Each jersey requires 10 balls of wool and a snood 4 balls of wool.

She will have to hire a van from time to time to deliver completed items to the distributors, and collect wool from a cash and carry. The best offer on van hire is this one.

Can you devise a business plan to advise Karen?

KEY QUESTIONS

3

K1 Solve this pair of simultaneous equations:

$2a - 3b = 12$
$2b + 3a = 5$

K2 On a cartesian diagram, show the region defined by the following set of inequations:

$x \geq 0$
$y \geq 0$
$x + y \leq 9$
$4x + y \leq 12$

K3 For the region defined in **K2**, state the maximum and minimum values of $x + y$.

INDICES AND ALGEBRAIC FRACTIONS
– THE SIMPLER THE BETTER

● My company is worth £50 million.
That is $£5 \times 10^7$.

$$-1 = \sqrt{-1} \times \sqrt{-1}$$
$$-1 = \sqrt{-1 \times -1}$$
$$-1 = \sqrt{1}$$
$$-1 = 1$$

That should
change the face
of mathematics

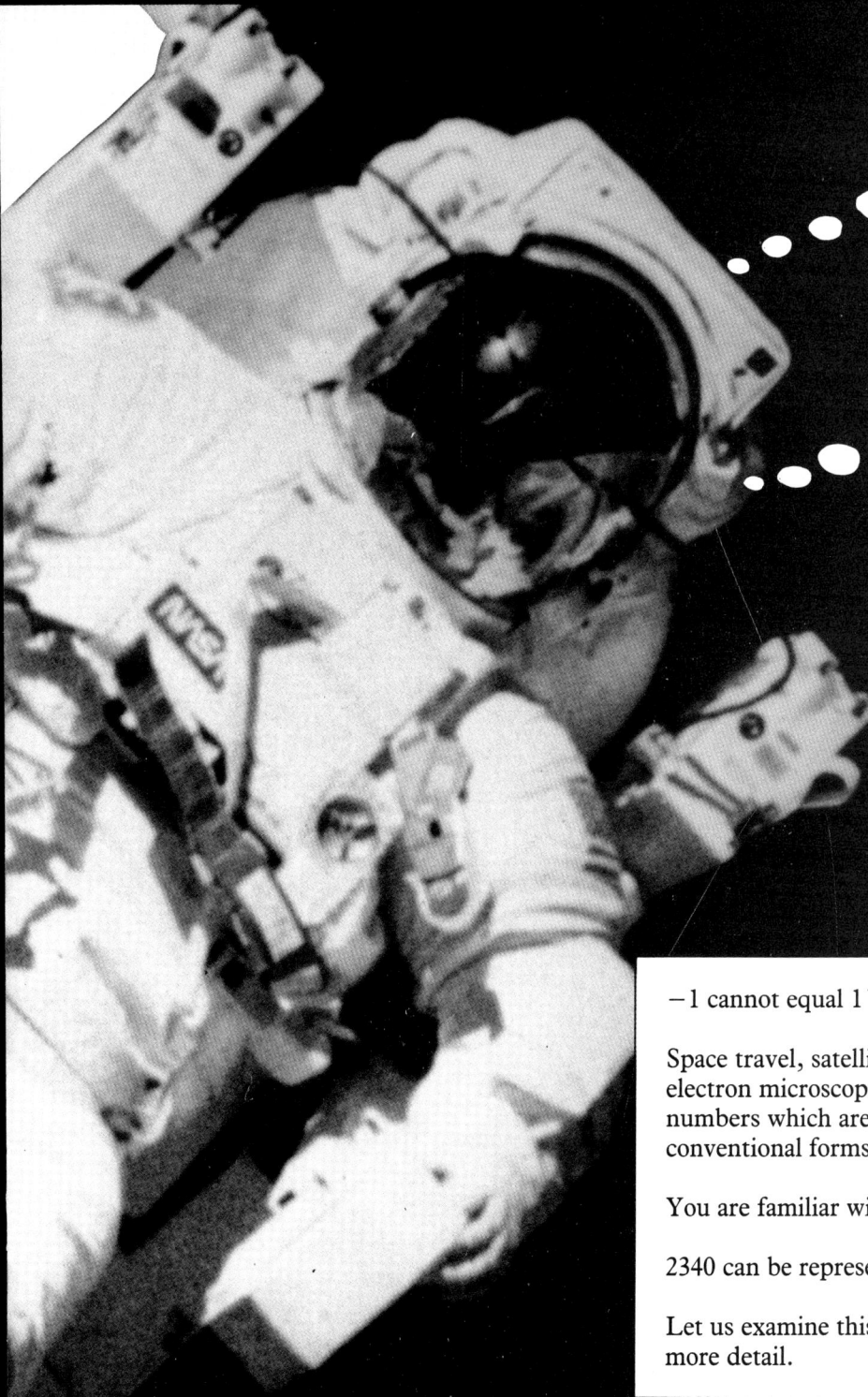

My speed is $2 \cdot 5 \times 10^5$ km/hr

That is 250 000 km/hr

−1 cannot equal 1?

Space travel, satellite television, astronomy, electron microscopy, high finance, all use numbers which are difficult to represent in conventional forms.

You are familiar with standard form.

2340 can be represented as $2 \cdot 34 \times 10^3$.

Let us examine this area of mathematics in more detail.

Rules of indices

3^6 is read as '3 to the power of 6'.

6 is called the **index** of the power.

3^6 is a shorthand way of writing $3 \times 3 \times 3 \times 3 \times 3 \times 3$.

Try these

1 (a) Complete the following:

$3^4 \times 3^3 = 3 \times 3 \times 3 \times 3 \times 3 \times 3 \times 3 = 3^?$

(b) Simplify (give your answer in index form)

$2^7 \times 2^5$

(c) Simplify (give your answer in index form)

$2^3 \times 3^4$

(d) Simplify $5^3 \times 5^x$

(e) Simplify $a^3 \times a^4$

(f) Simplify $a^m \times a^n$

Can you explain your answer?

2 (a) Complete the following:

$5^7 \div 5^4 = \dfrac{5 \times 5 \times 5 \times 5 \times 5 \times 5 \times 5}{5 \times 5 \times 5 \times 5} = \dfrac{5 \times 5 \times 5}{1} = 5^?$

(b) Simplify (giving your answer in index form)

$3^5 \div 3^3$

(c) Simplify (giving your answer in index form)

$5^4 \div 2^3$

(d) Simplify $a^6 \div a^2$

(e) Simplify $a^m \div a^n$

Can you explain your answer?

(f) Simplify $3^4 \div 3^x$

What happens if $x \geqslant 4$?

RULE 1

$$a^m \times a^n = a^{m+n}$$

When the number base is the same, you add the indices (plural of index).

RULE 2

$$a^m \div a^n = a^{m-n}$$

When the number base is the same, you subtract the indices.

We shall return at a later stage in this unit to examples when $n \geqslant m$.

Try this

3 Simplify the following (leaving answers in index form):

(a) $x^4 \times x^3$

(b) $b^3 \times b$

(c) $x^5 \div x^2$

(d) $a^3 \div a^2$

(e) $2x^2 \times x^3$

(f) $3a^3 \times 2a^4$

(g) $6x^7 \div 2x^5$

(h) $2m^3 \times 2m^3$

(i) $3x^3 \times 2y^2$

(j) $\dfrac{3^4 \times 3^3}{3^2}$

(k) $\dfrac{5^{10}}{5^4 \times 5^3}$

(l) $\dfrac{x^4 \times x^3}{x^5}$

(m) $\dfrac{2a^3 \times 3a^2}{6a}$

(n) $\dfrac{4p^2}{2p^2}$

(o) $\dfrac{2p^3(p+3)}{p^2}$

4

Let us return to rule 2: $a^m \div a^n = a^{m-n}$ and look at examples when $n = m$.

Consider

$x^5 \div x^5 = x^{5-5} = x^0$

$x^5 \div x^5 = \dfrac{x^5}{x^5} = \dfrac{x \times x \times x \times x \times x}{x \times x \times x \times x \times x} = \dfrac{1}{1} = 1$

So $x^0 = 1$

Likewise,

$a^m \div a^m = a^{m-m} = a^0$

$a^m \div a^m = \dfrac{a^m}{a^m} = 1$

RULE 3 $a^0 = 1$ (zero index)

Consider

$x^3 \div x^5 = x^{3-5} = x^{-2}$

$x^3 \div x^5 = \dfrac{x^3}{x^5} = \dfrac{x \times x \times x}{x \times x \times x \times x \times x} = \dfrac{1}{x^2}$

So $x^{-2} = \dfrac{1}{x^2}$

RULE 4 $a^{-m} = \dfrac{1}{a^m}$

Consider

$(x^2)^3 = x^2 \times x^2 \times x^2 = x^{2+2+2} = x^6$

So $(x^2)^3 = x^6$

RULE 5 $(a^m)^n = a^{mn}$

To find a power, you multiply the indices.

Try these

4 (a) 2^{-2} (b) $3^4 \div 3^4$ (c) $5^5 \times 5^{-3}$
   (d) $(2^2)^2$ (e) $3^2 \div 3^0$ (f) $3^2 \div 3^4$

5 Simplify (leaving your answers in index form)

(a) $x^3 \times x^5$ (b) $a^4 \times a^3 \times a$

(c) $b^7 \div b^5$ (d) $x^3 \div x^{-4}$

(e) $(a^3 \div a) \div a$ (f) $x^2(x^3 \div x^2)$

(g) $(2x)^2$ (h) $(a^2b)^3$

(i) $\dfrac{a^9 \times a}{a^2 \times a^3}$ (j) $\dfrac{x^2 \times x^4}{x^5 \times x}$

(k) $2x^3 \div x^2$ (l) $\left(\dfrac{x}{x^3}\right)^4$

(m) $3a^2b \times 2ab^2$ (n) $x(x^{-2} + x^3)$

Consider

$$a^{\frac{1}{2}} \times a^{\frac{1}{2}} = a^{\frac{1}{2}+\frac{1}{2}} = a \qquad\qquad a^{\frac{1}{3}} \times a^{\frac{1}{3}} \times a^{\frac{1}{3}} = a$$

So $(a^{\frac{1}{2}})^2 = a \qquad\qquad\qquad\qquad (a^{\frac{1}{3}})^3 = a$

$\qquad\quad a^{\frac{1}{2}} = \sqrt{a} \qquad\qquad\qquad\qquad\quad\ a^{\frac{1}{3}} = \sqrt[3]{a}$

(Note that $\sqrt[3]{\ }$ means cube root.)

So $\quad a^{\frac{1}{m}} = \sqrt[m]{a}$ (The m root of a)

Consider $\qquad a^{\frac{2}{3}} = (a^2)^{\frac{1}{3}} \quad$ or $\quad \sqrt[3]{a^2}$

$\qquad$ or $\qquad a^{\frac{2}{3}} = (a^{\frac{1}{3}})^2 \quad$ or $\quad (\sqrt[3]{a})^2$

RULE 6

$$a^{\frac{m}{n}} = \sqrt[n]{a^m} \qquad\quad \text{or} \qquad a^{\frac{m}{n}} = (\sqrt[n]{a})^m$$

Try these

6 Evaluate (give positive answer only).

(a) $\sqrt[3]{27}$ (b) $\sqrt[4]{81}$

(c) $36^{\frac{1}{2}}$ (d) $64^{\frac{1}{6}}$

(e) $49^{-\frac{1}{2}}$ (f) $16^{\frac{3}{4}}$

(g) $9^{\frac{3}{2}}$ (h) $27^{\frac{2}{3}}$

(i) $(\frac{1}{2})^{-2}$ (j) $(\frac{1}{5})^{-1}$

(k) $(16 \times 25)^{\frac{1}{2}}$ (l) $\sqrt[4]{160\ 000}$

7 Substitute $a=2$, $b=-1$ and $c=4$ in the following and evaluate:

(a) c^a (b) b^a (c) $3c^2b$ (d) $c^{\frac{1}{a}}$

8 Write the following as powers of 2:

(a) 4   (b) $\frac{1}{4}$ (c) 4^3 (d) $32^{\frac{1}{2}}$

9 Which is the largest, 3^{21}, 9^{10} or $(81)^4$?

Numbers in standard form

You met standard form in the *Red Book*. Standard form is often referred to as being scientific notation. A scientist working in the field of astronomy would not write down large numbers in conventional form. Let us revise some of the work you did in the *Red Book*.

Here is how a scientist would write down a number such as 124 000 000

$$124\ 000\ 000 = 1{\cdot}24 \times 100\ 000\ 000 = 1{\cdot}24 \times 10^8$$

Likewise, a scientist would not record small numbers such as 0·000243 but instead would do as follows:

$$0{\cdot}000243 = 2{\cdot}43 \times 0{\cdot}0001 = 2{\cdot}43 \times \frac{1}{10\ 000} = 2{\cdot}43 \times \frac{1}{10^4}$$

$$= 2{\cdot}43 \times 10^{-4}$$

A number is expressed in standard form when it is written in the form $a \times 10^n$ where a is a number between 1 and 10 and n is an integer (positive, negative or zero).

Try these

10 State the following numbers in standard form and in conventional (normal) form.

Note 1 billion is 10^9 (i.e. 1 000 000 000) and 1 million is 10^6.

(a) The defence budget is in excess of £3 billion.

(b) The trade deficit this month was £2·4 billion.

(c) The assets of this company are in excess of £300 million.

(d) The sun burns the equivalent of 1 tonne of hydrogen in approximately 0·000 000 001 seconds.

11 Calculate the following. Express answers in standard form.

(a) $(4 \times 10^3) \times (2 \times 10^2)$

(b) $(5 \times 10^4) \times (4 \times 10^2)$

(c) $(1·4 \times 10^{-3}) \times (4 \times 10^6)$

(d) $(2·4 \times 10^{-2}) \div (5 \times 10^4)$

Surds

Consider

$$\sqrt{4 \times 25} = \sqrt{100} = 10 \qquad \sqrt{4} \times \sqrt{25} = 2 \times 5 = 10$$
$$\text{So } \sqrt{4 \times 25} = \sqrt{4} \times \sqrt{25}$$

RULE 7 $\qquad \sqrt{ab} = \sqrt{a} \times \sqrt{b}$

Consider

$$\sqrt{\frac{36}{9}} = \sqrt{4} = 2 \qquad \frac{\sqrt{36}}{\sqrt{9}} = \frac{6}{3} = 2$$

RULE 8 $\qquad \sqrt{\dfrac{a}{b}} = \dfrac{\sqrt{a}}{\sqrt{b}}$

Rules 7 and 8 were illustrated by examples where only the positive value of a was considered. However, $\sqrt{4} = \pm 2$. Check that the results still hold for negative values of $\sqrt{\ }$.

$\sqrt{16}, \sqrt{\frac{9}{4}}, \sqrt{\frac{1}{25}}$ are rational numbers. Their answers ($\sqrt{16} = \pm 4$; ($\sqrt{\frac{9}{4}} = \pm\frac{3}{2}$; $\sqrt{\frac{1}{25}} = \pm\frac{1}{5}$) can be written in the form $\frac{p}{q}$ where p and q are integers.

Numbers like $\sqrt{2}$, $\sqrt{3}$, $\sqrt{5}$, $\sqrt{8}$ are called **surds**. These are square roots which cannot be evaluated precisely. There are methods of determining approximate answers to as many decimal places as required. The answers are never exact. Surds are members of the set of **irrational numbers**. The irrational numbers are included within the set of real numbers.

Try these

12 Which of the following numbers are surds?

(a) $\sqrt{36}$ (b) $\sqrt{7}$ (c) $\sqrt{\frac{16}{9}}$ (d) $\sqrt{8}$

(e) $\sqrt{0 \cdot 01}$ (f) $\sqrt[3]{27}$ (g) $\sqrt[3]{1}$ (h) $\sqrt[3]{-1}$

13 Simplify the following.
For example $\sqrt{45} = \sqrt{9 \times 5} = \sqrt{9} \times \sqrt{5} = 3\sqrt{5}$

(a) $\sqrt{8}$ (d) $\sqrt{20}$ (g) $\sqrt{300}$

(b) $\sqrt{32}$ (e) $\sqrt{\frac{8}{9}}$ (h) $\sqrt[3]{16}$

(c) $\sqrt{27}$ (f) $\sqrt{\frac{45}{25}}$ (i) $\sqrt[3]{-16}$

14 Simplify the following.
For example $\sqrt{3} \times \sqrt{6} = \sqrt{3} \times \sqrt{3} \times \sqrt{2} = 3\sqrt{2}$

(a) $\sqrt{3} \times \sqrt{3}$ (d) $\sqrt{2} \times \sqrt{18}$

(b) $\sqrt{6} \times \sqrt{2}$ (e) $\sqrt{10} \times \sqrt{40}$

(c) $\sqrt{5} \times \sqrt{5}$ (f) $\sqrt{3} \times \sqrt{27}$

15 Simplify the following (leave answers as surds).

(a) $\sqrt{12}$ (d) $\sqrt{3} \times \sqrt{300}$

(b) $\sqrt{2} \times \sqrt{50}$ (e) $2\sqrt{5} \times 3\sqrt{5}$

(c) $\sqrt{1000}$ (f) $\sqrt{5} \times \sqrt{15}$

Rationalising the denominator

4

By the definition of a square root, $\sqrt{3} \times \sqrt{3} = 3$; $\sqrt{5} \times \sqrt{5} = 5$.
So if there is a surd in the denominator (bottom) of a

fraction (e.g. $\dfrac{1}{\sqrt{2}}$), you can make the denominator rational by

multiplying top and bottom by the surd

(e.g. $\dfrac{1}{\sqrt{2}} = \dfrac{1}{\sqrt{2}} \times \dfrac{\sqrt{2}}{\sqrt{2}} = \dfrac{\sqrt{2}}{2}$).

For example, to rationalise the denominator of $\dfrac{10}{\sqrt{5}}$ multiply
top and bottom by $\sqrt{5}$:
$$\frac{10}{\sqrt{5}} = \frac{10}{\sqrt{5}} \times \frac{\sqrt{5}}{\sqrt{5}} = \frac{10 \times \sqrt{5}}{5} = 2\sqrt{5}$$

Try this

16 Rationalise the denominators of the following fractions and
simplify where possible.

(a) $\dfrac{1}{\sqrt{5}}$
(b) $\dfrac{2}{\sqrt{2}}$
(c) $\dfrac{3}{\sqrt{6}}$

(d) $\dfrac{3}{5\sqrt{12}}$
(e) $\dfrac{14}{2\sqrt{7}}$
(f) $\dfrac{5}{\sqrt{50}}$

Triangle ABC is an equilateral triangle of side 2 units.
AD is an axis of symmetry.

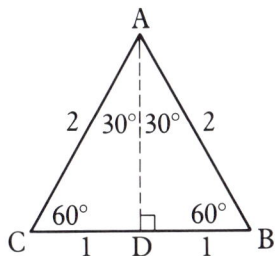

Triangle ABD is a right-angled triangle.
By using Pythagoras Theorem you can
calculate that $AD = \sqrt{3}$.

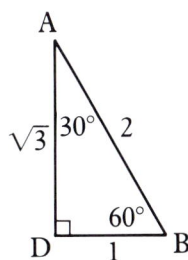

From this triangle you can state that $\sin 60° = \dfrac{\sqrt{3}}{2}$.

(Remember SOH CAH TOA from the *Yellow Book*.)

17 Express the following trig ratios as surds. Where required, rationalise the denominators.

(a) cos 60°   (b) tan 60°   (c) sin 30°
(d) tan 30°   (e) cos 30°

Algebraic fractions

$\dfrac{x}{2}$; $\dfrac{a^2b}{c}$ and $\dfrac{2y^2}{3x}$ are examples of algebraic fractions.

The rules for adding, subtracting, multiplying and dividing algebraic fractions are the same as those for numerical fractions. But the use of symbols (e.g. a, x, y, etc) can make them a little tricky to handle. So a bit of practice is helpful.

Multiplication

You will remember how to do this: $\frac{2}{3} \times \frac{4}{5}$.

$$\frac{2}{3} \times \frac{4}{5} = \frac{2 \times 4}{3 \times 5} = \frac{8}{15}$$

When calculating $\frac{1}{7} \times 5$, treat 5 as $\frac{5}{1}$.

So, $\frac{1}{7} \times 5 = \frac{1}{7} \times \frac{5}{1} = \frac{1 \times 5}{7 \times 1} = \frac{5}{7}$

Try this

18 Write each of the following as a single algebraic fraction:

(a) $\dfrac{x}{y} \times a$ (d) $\dfrac{2x^2}{y} \times \dfrac{x}{y}$

(b) $\dfrac{x}{y} \times \dfrac{x}{2}$ (e) $b^2 \times \dfrac{c}{a}$

(c) $\dfrac{a}{b} \times \dfrac{c}{b}$ (f) $\dfrac{a}{b} \times \dfrac{c}{d-e}$

Division

To divide by a fraction (e.g. $\frac{1}{2} \div \frac{2}{3}$) multiply by the fraction inverted (e.g. $\frac{1}{2} \times \frac{3}{2}$)

So $\frac{1}{2} \div \frac{2}{3} = \frac{1}{2} \times \frac{3}{2} = \frac{1 \times 3}{2 \times 2} = \frac{3}{4}$

Try this

4

19 Write each of the following as a single algebraic fraction.

(a) $\dfrac{p}{q} \div \dfrac{r}{s}$ (d) $\dfrac{2x^2}{y} \div \dfrac{y}{x}$

(b) $\dfrac{x}{y} \div \dfrac{2}{x}$ (e) $\dfrac{x}{y} \div a$

(c) $\dfrac{b^2}{a} \div \dfrac{c}{b}$ (f) $b^2 \div \dfrac{c}{a}$

Simplifying fractions

The fraction $\frac{16}{20}$ can be simplified to $\frac{4}{5}$.

$\frac{16}{20} = \frac{4 \times 4}{5 \times 4} = \frac{4}{5}$

Likewise, $\frac{25}{20}$ can be simplified to $1\frac{1}{4}$.

$\frac{25}{20} = \frac{5 \times 5}{4 \times 5} = \frac{5}{4} = 1\frac{1}{4}$

EXAMPLE

Simplify $\dfrac{2x^2y}{3xy^2}$.

$\dfrac{2x^2y}{3xy^2} = \dfrac{2 \times x \times x \times y}{3 \times x \times y \times y} = \dfrac{2 \times x}{3 \times y} = \dfrac{2x}{3y}$

EXAMPLE

Simplify $\dfrac{6x}{3x+9}$.

$\dfrac{6x}{3x+9} = \dfrac{\not{3} \times 2x}{\not{3}(x+3)} = \dfrac{2x}{x+3}$

Try this

20 Simplify these.

(a) $\dfrac{ab}{ac}$ (d) $\dfrac{5pq^2}{3q}$

(b) $\dfrac{4x}{2x^2}$ (e) $\dfrac{ab}{ax+ay}$

(c) $\dfrac{3ab}{2a}$ (f) $\dfrac{n^2+2n}{3n}$

Addition and subtraction

It is only possible to add or subtract fractions if the denominators are the same.

You will remember how to do this. E.g. $\frac{4}{5} - \frac{2}{3}$.

Make the denominator 15.

So, $\frac{4}{5} - \frac{2}{3} = \frac{4 \times 3}{5 \times 3} - \frac{2 \times 5}{3 \times 5} = \frac{12}{15} - \frac{10}{15} = \frac{2}{15}$.

EXAMPLE

Express $\frac{2}{b} - \frac{3}{a}$ as a single fraction.

Common denominator is ab.

$$\frac{2}{b} - \frac{3}{a} = \frac{2 \times a}{b \times a} - \frac{3 \times b}{a \times b}$$

$$= \frac{2a}{ab} - \frac{3b}{ab}$$

$$= \frac{2a - 3b}{ab}$$

EXAMPLE

Express $\frac{x}{3} + \frac{x}{x+2}$ as a single fraction.

Common denominator is $3(x+2)$.

$$\frac{x}{3} + \frac{x}{x+2} = \frac{x(x+2)}{3(x+2)} + \frac{x \times 3}{(x+2) \times 3}$$

$$= \frac{x^2 + 2x}{3(x+2)} + \frac{3x}{3(x+2)}$$

$$= \frac{x^2 + 2x + 3x}{3(x+2)}$$

$$= \frac{x^2 + 5x}{3x + 6}$$

Try this

21 Express each of these as a single fraction.

(a) $\frac{1}{a} + \frac{1}{b}$

(b) $\frac{a}{2} + \frac{1}{a}$

(c) $\frac{x}{y} - \frac{a}{b}$

(d) $\frac{x}{2} - \frac{2}{x}$

(e) $\frac{a}{4b} - \frac{1}{2ab}$

(f) $\frac{1}{x+1} + \frac{2}{x}$

(g) $\frac{y}{a} - \frac{y}{a+2}$

(h) $\frac{1}{a^2} - \frac{1}{a^2+1}$

(i) $\frac{x}{x-3} - \frac{x}{x+3}$

Equations with algebraic fractions

4

EXAMPLE

Solve the equation $\dfrac{4}{x}=\dfrac{2}{x+4}$.

Make the denominator the same [i.e. $x(x+4)$].

$$\frac{4}{x}=\frac{2}{x+4}$$

$$\frac{4(x+4)}{x(x+4)}=\frac{2x}{(x+4)x}$$

$$\frac{4x+16}{x(x+4)}=\frac{2x}{x(x+4)}$$

Now that the denominators are the same, you can equate the numerators:

$$4x+16=2x$$
$$4x-2x=-16$$
$$2x=-16$$
$$x=-8$$

Try this

22 Solve each of these equations.

(a) $\dfrac{x}{2}=x-4$

(b) $\dfrac{x-10}{3}=2x$

(c) $\dfrac{x+1}{3}=x$

(d) $\dfrac{x}{3}=\dfrac{x-2}{2}$

(e) $\dfrac{x-1}{2}=\dfrac{x-4}{5}$

(f) $\dfrac{4}{x-2}=\dfrac{8}{x}$

(g) $\dfrac{x}{4}=\dfrac{9}{x}$

(h) $\dfrac{a}{2}-\dfrac{9}{a}=0$

(i) $\dfrac{7}{x-3}=\dfrac{3}{x+1}$

KEY QUESTIONS

K1 Simplify

(a) $3x^2\times 2x^5$

(b) $8y^4\div 2y^3$

(c) $(x^2y)^3$

(d) $16^{-\frac{1}{2}}$

(e) $\left(\dfrac{1}{3}\right)^{-2}$

(f) $(3\times 10^4)\times(2\times 10^{-2})$

K2 Simplify

(a) $\sqrt{12}$

(b) $\sqrt{6}\times\sqrt{12}$

K3 Express each of the following as a single fraction.

(a) $\dfrac{a}{a-2}+\dfrac{a}{a+3}$

(b) $\dfrac{xy}{x^2y+xy^2}$

K4 Solve the equation

$$\frac{x}{5}=\frac{x-4}{4}$$

55

Now try this...

A This statement appeared at the start of the unit:

$$-1 = \sqrt{-1} \times \sqrt{-1}$$
$$-1 = \sqrt{-1 \times -1}$$
$$-1 = \sqrt{1}$$
$$-1 = 1$$

This is obviously not correct. Why?

B PERFECT NUMBERS

7 is a prime number. Prime numbers are those whole numbers which are only divisible by 1 or themselves. So the only factors of 7 are 1 and 7.

The factors of 10 are 1, 2, 5 and 10. The sum of the factors, apart from 10, is $1 + 2 + 5 = 8$ which is less than 10.

The factors of 12 are 1, 2, 3, 4, 6 and 12. The sum of the factors, apart from 12, is $1 + 2 + 3 + 4 + 6 = 16$ which is greater than 12.

A few numbers are equal to the sum of their factors, such as 6.

Factors of 6 are 1, 2, 3 and 6. The sum is $1 + 2 + 3 = 6$. These numbers are called **perfect numbers**.

1 See if you can find any other perfect numbers which are less than 40.
2 Euclid claimed that any number of the form $2^{n-1}(2^n - 1)$ is perfect so long as $2^n - 1$ is prime. Test this formula to see if it works.

C
$$2 + 2 = 2 \times 2$$
$$3 + 1\tfrac{1}{2} = 3 \times 1\tfrac{1}{2}$$
$$6 + 1 \cdot 2 = 6 \times 1 \cdot 2$$

An infinite number of pairs of numbers have the same sum and product. See if you can find some more.

Try to establish a method for generating pairs of numbers which have the same sum and product.

QUADRATICS

You should be able to make sense of the examples on this page after you have done the work of this unit

A league contains n teams.

Each team plays every other team twice—once at home, once away.

The number of matches is a quadratic expression

$$n(n-1) \qquad \text{or} \qquad n^2 - n$$

The temperature, T degrees centigrade, at which water boils is related to the height, H metres above sea level by the following quadratic expression

$$H = 1000(100 - T) + 580(100 - T)^2$$

According to Newton's law of gravitation, the magnitude of the gravitational force between two objects is given by the expression

$$g = \frac{Km_1m_2}{d^2}$$

EARTH

414000 km

46000 km

Moon

where m_1 and m_2 are the masses of the objects, d is the distance between them and K is a constant

This can be used to show, for example, that an object on a line between the Earth and the Moon (which are $4{\cdot}60 \times 10^8$ m apart) will experience the same gravitational pull from each planet when it is $4{\cdot}14 \times 10^8$ m from the Earth.

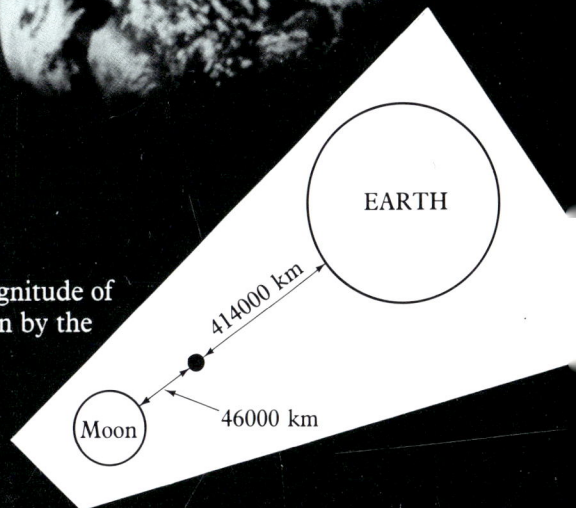

In the *Now try this* section of an earlier unit in this book you were asked to think about the **zeros** and **minimum** of the expression $x^2 - 2x - 3$.

(The **zeros** are the values of x which give the expression the value zero; the **minimum** is the value of x which gives the expression its minimum value.)

We put some values in a table:

x	$x^2 - 2x - 3$
.	.
.	.
.	.
4	5
3	0
2	-3
1	-4
0	-3
-1	0
-2	5

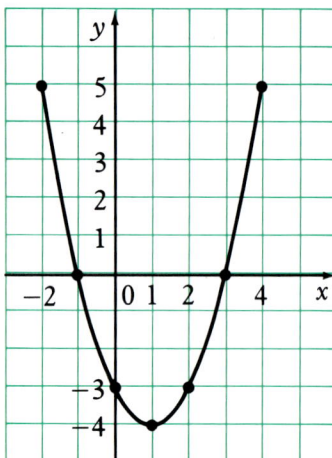

and suggested you draw a graph of the function $y = x^2 - 2x - 3$, $x \in R$.

From the table, we see that

$x = 3$ is a zero of $x^2 - 2x - 3$
$x = -1$ is a zero of $x^2 - 2x - 3$

and from the graph, it seems reasonable to say that the minimum value of $x^2 - 2x - 3$ is -4 and occurs when $x = 1$.

You were asked whether finding the **factors** of $x^2 - 2x - 3$ would help to find its zeros and minimum.

Let's do that now:.

$x^2 - 2x - 3 = (x - 3)(x + 1)$ (See unit 1.)

The factors of $x^2 - 2x - 3$ are $x - 3$ and $x + 1$.

The **zeros** of $x^2 - 2x - 3$ are the values of x that make $x^2 - 2x - 3$ zero.

The values of x that make $x^2 - 2x - 3$ zero are the values of x which make the factors of $x^2 - 2x - 3$ zero.

$x - 3 = 0$ or $x + 1 = 0$
     $x = 3$ or $x = -1$

These are the zeros which we found from our table of values. **Factorising the expression is a quick way of finding its zeros**.

From the graph, we see that the minimum is half-way between the zeros. Half-way between $x = -1$ and $x = 3$ is $x = 1$; and when $x = 1$, $x^2 - 2x - 3 = -4$, the minimum value.

Here are some more examples

expression	factors	zeros	minimum	minimum value
$x^2 - 6x + 8$	$(x - 2), (x - 4)$	$x = 2, x = 4$	$x = 3$	-1
$x^2 + 10x + 21$	$(x + 7), (x + 3)$	$x = -7, x = -3$	$x = -5$	-4
$x^2 - 4x - 5$	$(x - 5), (x + 1)$	$x = 5, x = -1$	$x = 2$	-9

Try this

1 Complete this table

expression	factors	zeros	minimum	minimum value
$x^2 + x - 2$				
$x^2 - x - 6$				
$x^2 - 4x + 3$				
$x^2 + 5x + 6$				
$x^2 - 3x - 10$				
$x^2 - 5x - 24$				

Let's consider now the expression $8 - 2x - x^2$.

Here is a table of values

x	$8 - 2x - x^2$
.	.
.	.
.	.
-5	-7
-4	0
-3	5
-2	8
-1	9
0	8
1	5
2	0
3	-7

and graph of the function $y = 8 - 2x - x^2$, $x \in R$.

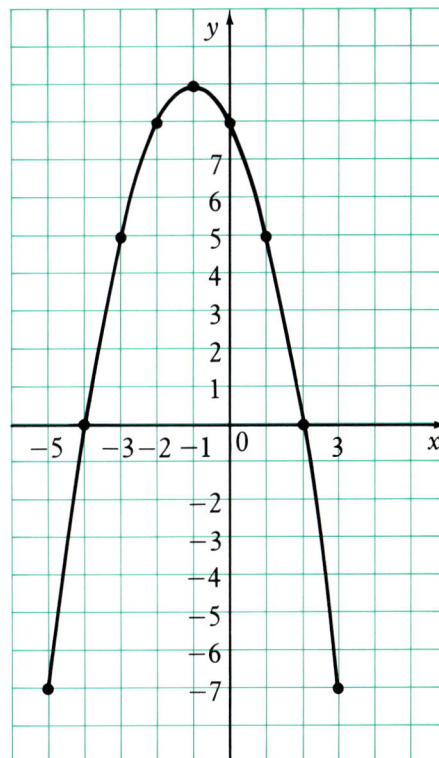

From the table and graph, the zeros are $x = -4$ and $x = 2$.

This is as we would have expected, since factorising gives
$8 - 2x - x^2 = (4 + x)(2 - x)$
$8 - 2x - x^2$ is zero only if one of its factors is zero,
i.e. only if $\quad 4 + x = 0 \quad$ or $\quad 2 - x = 0$
i.e. only if $\qquad x = -4 \quad$ or $\qquad x = 2$

The table and graph, however, show that we are looking at a **maximum** and not a minimum.

$x = -1$ is the maximum, giving $8 - 2x - x^2$ a maximum value of 9. (As before, the maximum is half-way between the zeros.)

Try this

2 Find the zeros, maximum and maximum value for each of these expressions

(a) $3 - 2x - x^2$ $\qquad$ (b) $3 + 2x - x^2$ $\qquad$ (c) $6 - x - x^2$

Finding the zeros of the expression $x^2 - 2x - 3$
is the same as
finding the roots of the equation $x^2 - 2x - 3 = 0,\ x \in R$
or
solving the equation.
($x \in R$ means that the roots of the equation belong to R, the set of real numbers.)

The equation $x^2 - 2x - 3 = 0$ is called a **quadratic** equation because x^2 (but no higher power of x) is in the equation.

$2x + 3 = 4 \quad$ and $\quad x^3 + 2x^2 - 3x + 4 = 0 \quad$ *are not* quadratic equations
$2x^2 + 5x - 12 = 0 \quad$ and $\quad 10x - x^2 = 0$ *are* quadratic equations.

Let's solve these quadratic equations
$2x^2 + 5x - 12 = 0,\ x \in R$
$(2x - 3)(x + 4) = 0$ (factorising $2x^2 + 5x - 12$; see unit 1)
$2x - 3 = 0 \quad$ or $\quad x + 4 = 0$
$\qquad x = \frac{3}{2} \quad$ or $\qquad x = -4$

The roots of the equation $2x^2 + 5x - 12 = 0,\ x \in R$, are $\frac{3}{2}$ and -4.

The minimum lies half-way between the roots, i.e. at $-\frac{5}{4}$.

The value of $2x^2 + 5x - 12$ when $x = -\frac{5}{4}$ (or $-1 \cdot 25$) is $-15 \cdot 125$.

So, the minimum value of $2x^2 + 5x - 12$ is $-15 \cdot 125$.
(You should check this using your calculator.)

$10x - x^2 = 0, \quad x \in R$
$x(10 - x) = 0$
$x = 0 \quad$ or $\quad 10 - x = 0$
$x = 0 \quad$ or $\qquad x = 10$

The roots of the equation $10x - x^2 = 0$, $x \in R$, are 0 and 10.
The maximum is when $x = 5$.
The maximum value of $10x - x^2$ is 25.

EXAMPLE

Solve the quadratic equation $50 + 5x - 6x^2$ and find its maximum value.
$$50 + 5x - 6x^2 = (10 - 3x)(5 + 2x)$$

The factors of $50 + 5x - 6x^2$ are $10 - 3x$ and $5 + 2x$.

$$50 + 5x - 6x^2 = 0$$
$$(10 - 3x)(5 + 2x) = 0$$
$$10 - 3x = 0 \quad \text{or} \quad 5 + 2x = 0$$
$$x = \tfrac{10}{3} \quad \text{or} \quad x = -\tfrac{5}{2}$$

The roots of $50 + 5x - 6x^2 = 0$ are $\tfrac{10}{3}$ and $-\tfrac{5}{2}$;

maximum lies half-way between roots, i.e. at $\tfrac{5}{12}$.

maximum value is $50 + 5 \times \left(\tfrac{5}{12}\right) - 6 \times \left(\tfrac{5}{12}\right)^2$

$$= 50 + 5 \times 0.41666\ldots - 6 \times (0.41666\ldots)^2$$
$$= 50 + 2.08333\ldots - 6 \times 0.1736111\ldots$$
$$= 50 + 2.08333\ldots - 1.041666\ldots$$
$$= 51.041666\ldots$$

The maximum value of $50 + 5x - 6x^2$ is 51.0 (to 3 significant figures) when $x = \tfrac{5}{12}$.

Try this

3 In each of the following, factorise the quadratic expression; solve the quadratic equation on R, the set of real numbers; calculate the maximum or minimum value of the expression and state the value of x which gives this maximum or minimum.

(a) $x^2 - 2x - 15 = 0$ (d) $x^2 + 5x = 0$

(b) $25 - 5x - 2x^2 = 0$ (e) $x - x^2 = 0$

(c) $12x^2 - 7x - 10 = 0$

EXAMPLE

Consider the following problem.

A gardener wishes to fence off a rectangular plot in the corner of his walled garden. He has 10 metres of fencing and wants to use all of it. He wants his plot to have an area of 16 metres squared.
What are the length and width of the rectangle?

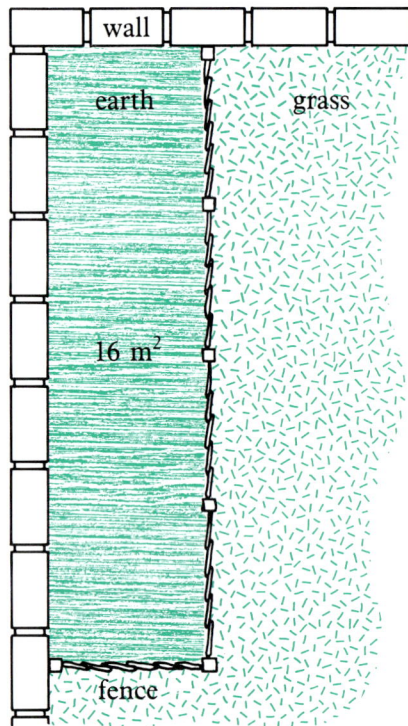

Let the length of the rectangle be x metres.
The width of the rectangle is $(10 - x)$ metres.
The area of the plot is $x \times (10 - x)$ metres squared.
The quadratic equation is $x(10 - x) = 16$.

Previous quadratic equations in this unit have been of the form

$$\textit{quadratic expression} = 0$$

We must put the quadratic equation for the gardening problem into that form before we can solve it.

The quadratic equation is
$$x(10 - x) = 16$$
$$10x - x^2 = 16$$
$$-x^2 + 10x - 16 = 0$$
$$x^2 - 10x + 16 = 0$$
$$(x - 8)(x - 2) = 0$$

The roots of the equation are $x = 8$ or $x = 2$.

If $x = 8$, the length is $8\,\text{m}$ and the width is $2\,\text{m}$ $(10 - x)$.
If $x = 2$, the length is $2\,\text{m}$ and the width is $8\,\text{m}$ $(10 - x)$.

Using either result, we can say that the sides of the rectangle are 8 metres and 2 metres.

You could, of course, have written down these answers without forming and solving a quadratic equation (and you can do so in the exercise that follows), but you will need to form quadratic equations to solve similar problems that will occur later in the unit.

Try these

4 Calculate the sides of the rectangle in the above gardening problem if

(a) there are 10 m of fencing and the area of the plot is 9 m²,

(b) there are 10 m of fencing and the area of the plot is 21 m²,

(c) there are 5 m of fencing and the area of the plot is 4 m²,

(d) there are 20 m of fencing and the area of the plot is 100 m².

5 Solve each of the following quadratic equations on R, the real numbers. (The first one has been started for you.)

(a) $x(x+1)=12$
$x^2+x=12$
$x^2+x-12=0$
. . . .

(b) $x(x-3)=10$

(c) $4x-x^2=-5$

(d) $x(9-x)=14$

(e) $3x(x+2)=(x+1)(x+6)$

(f) $(x-2)(x-9)=2x(x-4)$

Consider now the gardening problem when there are 10 metres of fencing and the area of the plot is 6 m².

The quadratic equation is
$$x(10-x)=6$$
$$10x-x^2=6$$
$$-x^2+10x-6=0$$
$$x^2-10x+6=0$$

The difficulty here is that $x^2-10x+6$ cannot be factorised using the same method, and we must use other methods to find the roots of the equation.

Iterative (repetitive) methods of solving quadratic equations

Let's rearrange the equation (yet again!)
$x^2-10x=-6$

Let us now draw a graph of the function
$y=x^2-10x \quad x \in R$

Table of values

x	x^2-10x	(x, y)
-1	11	$(-1, 11)$
0	0	$(0, 0)$
2	-16	$(2, -16)$
4	-24	$(4, -24)$
6	-24	$(6, -24)$
8	-16	$(8, -16)$
10	0	$(10, 0)$
11	11	$(11, 11)$

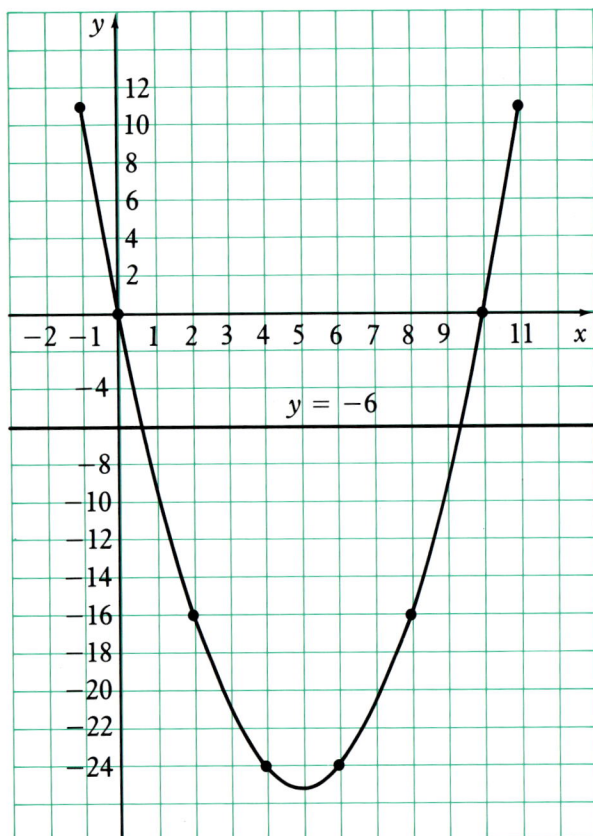

We can now find the roots of the equation
$x^2 - 10x = -6$
by drawing the straight line with equation $y = -6$ on the same diagram.

We see that the straight line cuts the curve at two points:
between $x = 0$ and $x = 1$ (at about $x = 0.6$) and
between $x = 9$ and $x = 10$ (at about $x = 9.4$).

We can make a better approximation to the 0.6 root by calculating the coordinates of more points between 0 and 1 and drawing the graph with a bigger scale between these points.

Table of values

x	$x^2 - 10x$	(x, y)
0	0	(0,0)
0·2	−1·96	(0·2, −1·96)
0·4	−3·84	(0·4, −3·84)
0·6	−5·64	(0·6, −5·64)
0·8	−7·36	(0·8, −7·36)
1	−9	(1, −9)

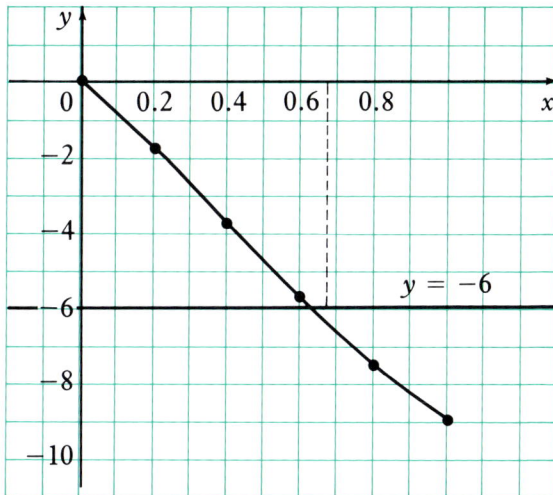

We see that the straight line $y = -6$ cuts the curve between 0·6 and 0·7 at about 0·65.

We can get a better approximation yet to the 0·65 root by repeating the graphing process
or
we can calculate the value of $x^2 - 10x$ for values of x round about 0·65 until we get a value of $x^2 - 10x$ as close to -6 as we want.

$x = 0.63 \qquad x^2 - 10x = -5.9031$
$x = 0.64 \qquad x^2 - 10x = -5.9904$
$x = 0.65 \qquad x^2 - 10x = -6.0775$
$x = 0.66 \qquad x^2 - 10x = -6.1644$

$x = 0.64$ is the best approximation to 2 decimal places.

One of the roots of the equation $x^2 - 10x = -6$, $x \in R$, is 0·64 to 2 decimal places.

Try this

6 Find, to 2 decimal places, the other root of the equation $x^2 - 10x = -6$, $x \in R$.

A formula for solving quadratic equations

We can obtain the roots of the quadratic equation
$x^2 - 10x = -6$
without using the above iterative processes.

This method involves writing $x^2 - 10x$ in the form
$(x + a)^2 - b$.

The process is called **completing the square** on the quadratic expression and it may help you to understand how we obtain the formula that follows. If you don't understand the method, don't worry. You can just use the formula instead!

$(x + a)^2 = x^2 + 2ax + a^2$

If $2ax$ has to be $-10x$, then $2a$ has to be -10 and a has to be -5.

$(x - 5)^2 = x^2 - 10x + 25$

If $(x - 5)^2 - b$ has to be $x^2 - 10x$, b has to be 25.

$x^2 - 10x = (x - 5)^2 - 25$

Now the equation $x^2 - 10x = -6$ can be written in the form

$(x - 5)^2 - 25 = -6$

$(x - 5)^2 = -6 + 25$
$(x - 5)^2 = 19$
$(x - 5) = \sqrt{19}$ or $-\sqrt{19}$
$x - 5 = 4 \cdot 36$ or $-4 \cdot 36$
$x = 5 + 4 \cdot 36$ or $5 - 4 \cdot 36$
$\quad = 9 \cdot 36$ or $0 \cdot 64$

The roots of the quadratic equation
$x^2 - 10x = -6 \quad x \in R$
are $0 \cdot 64$ and $9 \cdot 36$ (to 2 decimal places), as we found earlier by iterative methods.

Try this

7 Try using the above method to solve the following equations on R. If you can't, don't worry. You can use the formula in the next exercise

(a) $x^2 - 10x = -15$
(b) $x^2 - 11x = -5$
(c) $x^2 - 6x = 3$

To save having to go through the above process every time, we have a formula for solving quadratic equations.

The roots of the quadratic equation $ax^2 + bx + c = 0$
are given by

$$x = \frac{-b + \sqrt{(b^2 - 4ac)}}{2a} \quad \text{or} \quad x = \frac{-b - \sqrt{(b^2 - 4ac)}}{2a}$$

Let's check this with the equation $x^2 - 10x = -6$
$x^2 - 10x = -6$
$x^2 - 10x + 6 = 0$ and, comparing with $ax^2 + bx + c = 0$,
$a = 1$, $b = -10$, $c = 6$.

$$x = \frac{-(-10) + \sqrt{[(-10)^2 - 4 \times 1 \times 6]}}{2 \times 1} \quad \text{or} \quad x = \frac{-(-10) - \sqrt{[(-10)^2 - 4 \times 1 \times 6]}}{2 \times 1}$$

5

$$x = \frac{10 + \sqrt{(100 - 24)}}{2} \qquad \text{or} \qquad x = \frac{10 - \sqrt{(100 - 24)}}{2}$$

$$x = \frac{10 + \sqrt{76}}{2} \qquad \text{or} \qquad x = \frac{10 - \sqrt{76}}{2}$$

$$x = \frac{10 + 8{\cdot}718}{2} \text{ (3 decimal places)} \qquad \text{or} \qquad x = \frac{10 - 8{\cdot}718}{2} \text{ (3 decimal places)}$$

$$x = \frac{18{\cdot}718}{2} \qquad \text{or} \qquad x = \frac{1{\cdot}282}{2}$$

$$x = 9{\cdot}36 \text{ (2 decimal places)} \qquad \text{or} \qquad x = 0{\cdot}64 \text{ (2 decimal places)}$$

The roots of the equation $x^2 - 10x = -6$ are $0{\cdot}64$ and $9{\cdot}36$ (to 2 decimal places).

Try this

8 Use the quadratic formula to solve the three equations in *Try this 7*.

It is important, when using the quadratic formula, to rewrite the equation, if necessary, in the form $ax^2 + bx + c = 0$.

equation	rewritten equation	a, b, c
$x^2 - 10x = -6$	$x^2 - 10x + 6 = 0$	$a = 1$, $b = -10$, $c = 6$
$x^2 = 6x - 7$	$x^2 - 6x + 7 = 0$	$a = 1$, $b = -6$, $c = 7$
$15 - 5x - 10x^2 = 0$	$-10x^2 - 5x + 15 = 0$	$a = -10$, $b = -5$, $c = 15$

Try these

9 Rewrite each of the following equations in the form $ax^2 + bx + c = 0$ and write down the values of a, b and c for each equation.

(a) $x(x + 1) = 13$
(b) $x(x - 3) = 9$
(c) $4x - x^2 = -7$
(d) $x(9 - x) = 13$
(e) $(x - 3)(x - 8) = 2x(x - 5)$
(f) $3x(x - 2) = (x - 2)(x + 5)$

10 Solve each of the equations $9(a) - (f)$ on R using the quadratic formula giving your answers to 2 decimal places where necessary.

In one case, the square root, $\sqrt{b^2 - 4ac}$, is exact. This means that you could have solved the equation by factorising — perhaps you could try to find the factors.

In any case, it is comforting to know that we can use the formula to find roots of quadratic equations whether or not there are factors (and whether or not we can find them!)

11 For each of the following problems, form a quadratic equation, find its roots and write down the solution to the problem.

> *Note:* You will usually find two roots of the equation. It may be that both roots will be solutions to the problem, or perhaps only one of them makes sense. E.g. $x = 2$ makes sense as the length of a rectangle, but $x = -3$ does not. In each case, you must decide which of the roots are solutions to the problem.

(a) The perimeter of a rectangle is 40 cm and its area is 90 cm². What are the dimensions of the rectangle (to 1 decimal place)?

(b) The sum of the first n odd numbers, beginning at 1, is n^2. How many odd numbers are needed to add up to 121?

(c) The sum of the first n even numbers, beginning at 2, is $n(n+1)$. How many even numbers are needed to add up to 210?

(d) The hypotenuse of a right-angled triangle is 3 cm longer than the shortest side and 2 cm longer than the other side. Calculate the lengths of the sides of the triangle to 3 significant figures.

> *A reminder — the Theorem of Pythagoras*
> In a right-angled triangle, the hypotenuse squared is equal to the sum of the squares of the other two sides.

Quadratic equations — with fractions!

In an earlier unit in this book we looked at equations involving fractions. We will now extend this to quadratics.

Solve the equation

$$\frac{1}{x} + \frac{2}{2x-1} = \frac{7}{6} \quad x \in R$$

We multiply each term by $x \times (2x-1) \times 6$ (i.e. all the denominators multiplied together).

$$\frac{1 \times x \times (2x-1) \times 6}{x} + \frac{2 \times x \times (2x-1) \times 6}{(2x-1)} = \frac{7 \times x \times (2x-1) \times 6}{6}$$

and cancel within each fraction in the equation

$$\frac{1 \times \cancel{x} \times (2x-1) \times 6}{\cancel{x}} + \frac{2 \times x \times \cancel{(2x-1)} \times 6}{\cancel{(2x-1)}} = \frac{7 \times x \times (2x-1) \times \cancel{6}}{\cancel{6}}$$

giving

$$1 \times (2x-1) \times 6 + 2 \times x \times 6 = 7 \times x \times (2x-1)$$
$$12x - 6 \quad + \quad 12x \quad = 14x^2 - 7x$$
$$-14x^2 + 24x + 7x - 6 = 0$$
$$-14x^2 + 31x - 6 = 0$$
$$14x^2 - 31x + 6 = 0$$
$$(x-2)(14x-3) = 0$$
$$x - 2 = 0 \quad \text{or} \quad 14x - 3 = 0$$
$$x = 2 \quad \text{or} \quad x = \frac{3}{14}$$

The roots of the equation $\dfrac{1}{x} + \dfrac{2}{2x-1} = \dfrac{7}{6}$ are 2 and $\frac{3}{14}$.

Try this

12 Solve each of the following equations on R, giving the roots to 3 significant figures where necessary.

(a) $\dfrac{3}{x+3} - \dfrac{1}{3x-1} = \dfrac{1}{4}$ (b) $\dfrac{1}{x-1} + \dfrac{1}{x+1} = 1$ (c) $\dfrac{1}{x-1} - \dfrac{1}{x+1} = 1$

Now try this...

Consider the quadratic formulae

$$x = \frac{-b + \sqrt{(b^2 - 4ac)}}{2a} \quad \text{or} \quad x = \frac{-b - \sqrt{(b^2 - 4ac)}}{2a}$$

Since a negative number has no real square root, the quadratic equation has no real roots if $b^2 - 4ac$ is negative.

Let's look at such an example:
$x^2 + 6x + 10 = 0$
$a = 1, \ b = 6, \ c = 10$
$b^2 - 4ac = 6^2 - 4 \times 1 \times 10$
$\qquad = 36 - 40$
$\qquad = -4$

We cannot find roots for this quadratic equation.
We can see from the graph that the curve does not cut the x-axis in any point and this is more evidence that real roots do not exist.

We cannot therefore calculate the maximum or minimum of the quadratic expression in the way in which we have done in this unit so far — that is, by taking the max/min as half-way between the roots.

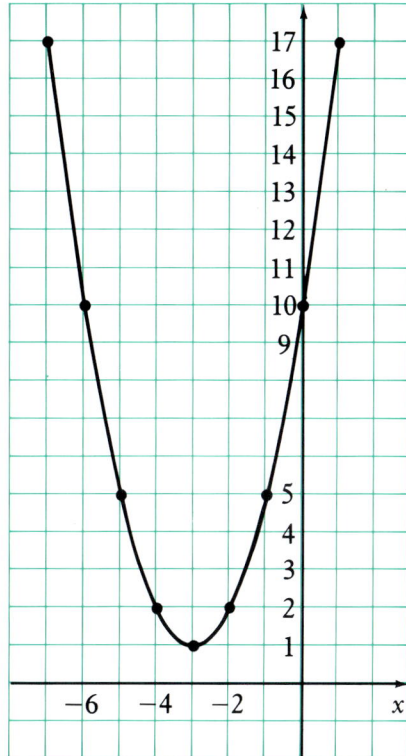

When we introduced the formula for solving quadratic equations we looked at a process called **completing the square** on a quadratic expression. This process will be useful to us now.

$x^2 + 6x + 10$
$x^2 + 6x$ has now to be written as $(x + a)^2 - b$
$x^2 + 6x = (x + 3)^2 - 9$

The expression now becomes
$[(x + 3)^2 - 9] + 10 = (x + 3)^2 + 1$

Can you now say for which value of x the expression $x^2 + 6x + 10$ has its minimum value and calculate this value?

Can you say for which value of x the expression $x^2 + 6x + 11$ has it minimum value and calculate this value?

Can you say what is the largest value c can take in the quadratic equation $x^2 + 6x + c = 0$
if the equation is to have real roots?

Can you say what is the largest value c can take in the quadratic equation $x^2 + 8x + c = 0$
if the equation is to have real roots?

KEY QUESTIONS

K1 Solve the equation
$3x^2 - 2x - 8 = 0 \quad x \in R$

K2 Calculate the maximum value of the quadratic expression
$3 + 2x - x^2$.

K3 Solve the equation
$3x^2 - 2x - 7 = 0 \quad x \in R$
giving your answers to 2 decimal places.

K4 Solve the equation
$3x(x-4) = (x-2)(x+3) \quad x \in R$.

K5 Solve the quadratic equation
$$\frac{1}{2x+1} - \frac{2}{x-1} = \frac{1}{6} \quad x \in R$$
giving your answers to 2 decimal places.

FUNCTIONS

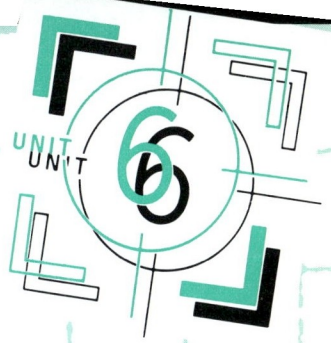

In this unit we will consider some more *functions*

You may remember that, in an earlier book, we described a function as a relation between two sets of numbers.

We drew graphs of some of them.

The ancient Greek mathematician Apollonius investigated the curves that are obtained by cutting through a cone.

These *conic* curves and their graphs are shown here.

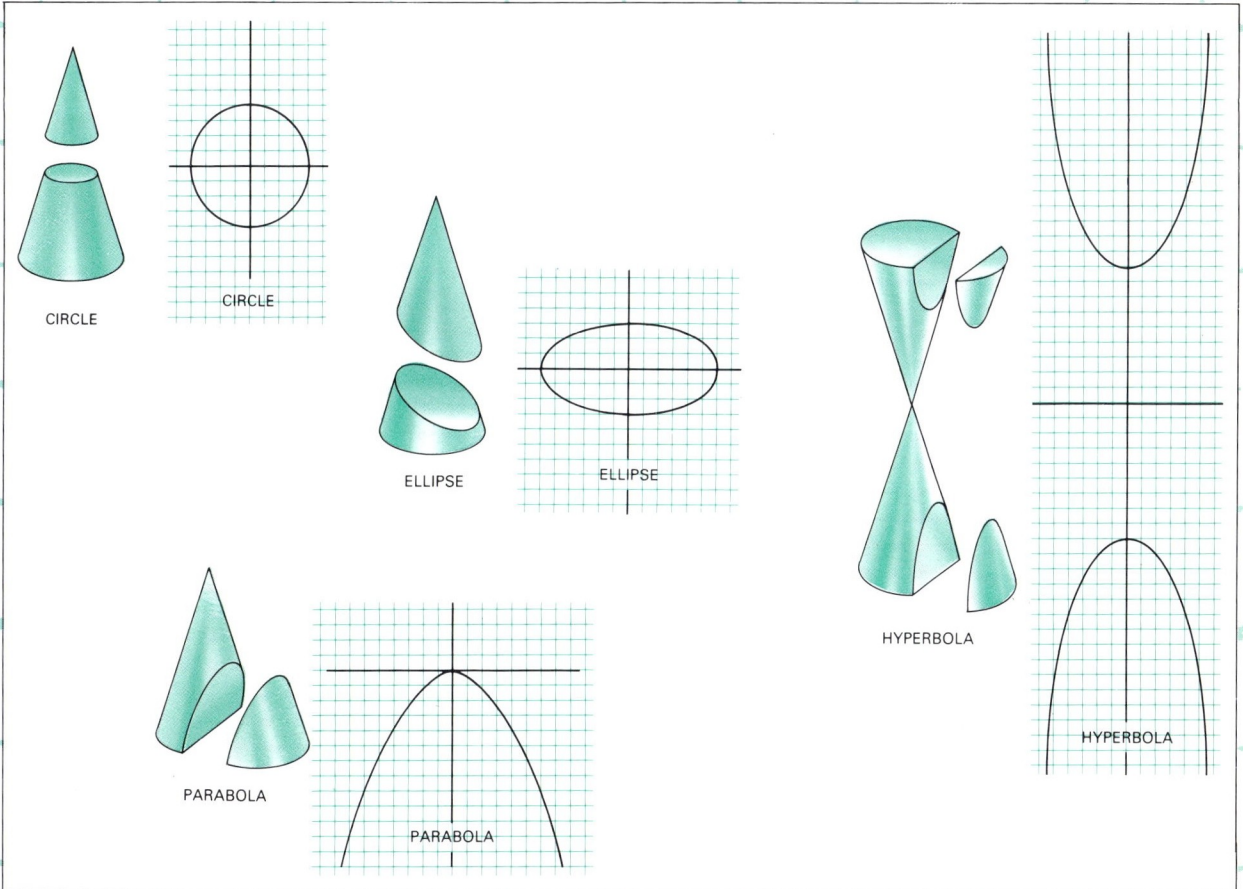

CIRCLE

CIRCLE

ELLIPSE

ELLIPSE

HYPERBOLA

HYPERBOLA

PARABOLA

PARABOLA

The first function (the quadratic) which we consider in this unit is related to one of the conic curves — the parabola.

You may have the opportunity to study some of the other conic curves if you take your study of mathematics further.

You may recall from one of our earlier books the **function**
$y = x^2 + x - 2 \quad x \in R$.

This is an example of a **quadratic function**.

We now introduce a new notation for functions:
$f(x) = x^2 + x - 2 \quad x \in R$

Here is a table of values for $f(x)$ and a graph of the function.

x	$x^2 + x - 2$	$(x, f(x))$
2	4	(2,4)
1	0	(1,0)
0	−2	(0, −2)
−1	−2	(−1, −2)
−2	0	(−2,0)
−3	4	(−3,4)

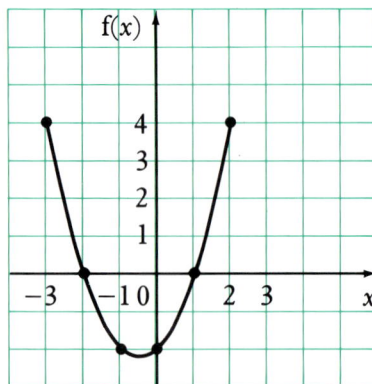

From earlier chapters in this book, you should know that
$x^2 + x - 2$
can be written as $(x + 2)(x - 1)$.

$x = -2$ and $x = 1$
are the **roots** or **zeros** of the equation
$x^2 + x - 2 = 0$.

Try this

1 (a) Consider another function, $g(x)$, where
$$g(x) = -f(x) \qquad x \in R$$
$$= -(x^2 + x - 2)$$
$$= -x^2 - x + 2$$
$$= 2 - x - x^2$$
Sketch a graph of $g(x)$.

(b) Consider now the function, $h(x)$, where
$h(x) = 2f(x) \qquad x \in R$
Sketch a graph of $h(x)$.

Note: We have used the letters f, g and h to label different functions and, in fact, we could use any letter to do so.

Where we are referring to only one function in a question, we will use $f(x)$ from now on.

Intervals on the real number line

In some parts of this unit, we will look at **intervals** on the
real number line e.g.
"$-3 < x < 2 \quad x \in R$" means
"real numbers from -3 to 2, not including -3 and not
including 2".

This interval can be shown in the following way

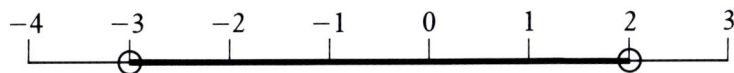

"$-4 \leqslant x \leqslant 1 \quad x \in R$" means
"real numbers from -4 to 1, including -4 and including 1"

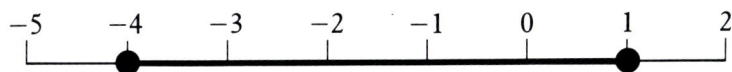

"$0 < x \leqslant 5 \quad x \in R$" means
"real numbers from 0 to 5, not including 0 but including 5".

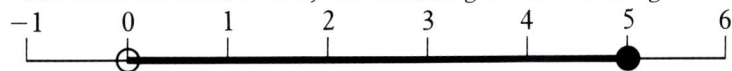

Try this

2 Complete this table

interval	number line
$-1 < x < 3 \qquad x \in R$	
$1 \leqslant x \leqslant 4 \qquad x \in R$	
$-4 < x \leqslant 0 \qquad x \in R$	

Now consider the function

$$f(x) = x^3 + 2x^2 - 5x - 6 \qquad x \in R$$

$x = 3$
$$f(3) = 3^3 + (2 \times 3^2) - (5 \times 3) - 6$$
$$= 27 + (2 \times 9) - \quad 15 \quad - 6$$
$$= 27 + \quad 18 \quad - \quad 15 \quad - 6$$
$$= 24$$

$x = -2$
$$f(-2) = (-2)^3 + [2 \times (-2)^2] - (5 \times -2) - 6$$
$$= -8 + \quad (2 \times 4) \quad - \quad (-10) \quad -6$$
$$= -8 + \quad\quad 8 \quad\quad + \quad 10 \quad -6$$
$$= \quad 4$$

Copy and complete this table

x	$x^3 + 2x^2 - 5x - 6$	$(x, f(x))$
3	24	(3,24)
2		
1		
0		
−1		
−2	4	(−2,4)
−3		
−4		

and then copy and complete the graph.

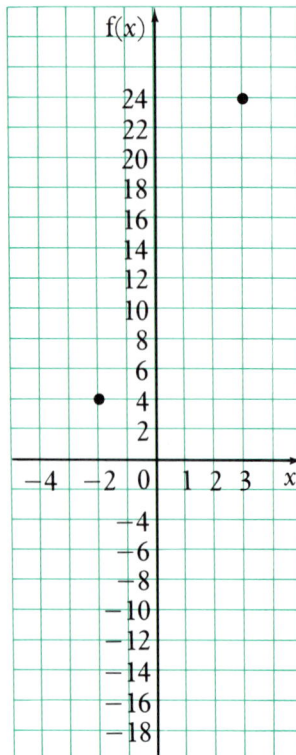

What are the roots of the equation
$$x^3 + 2x^2 - 5x - 6 = 0?$$

Try this

3 Make a table and then draw a graph for the function
$f(x) = x^3 - 2x^2 - 5x + 6$ $-3 \leqslant x \leqslant 4$ $x \in R$

The real numbers
from −3 to 4
including −3 and 4

What are the roots of the equation
$x^3 - 2x^2 - 5x + 6 = 0$?

In one of our earlier books, we looked at the function

$f(x) = \dfrac{10}{x}$ $x > 0$ $x \in R$
       The real numbers
       more than 0

and obtained the following table

x	$\dfrac{10}{x}$	(x, y)
1	10	(1,10)
2	5	(2,5)
4	2·5	(4,2·5)
5	2	(5,2)
10	1	(10,1)

Note: Since $\frac{10}{0}$ cannot be calculated, $f(x)$ is not defined at $x = 0$.

If we wish to extend $f(x)$ to all values of x except zero, we write

$f(x) = \dfrac{10}{x}$ $x \neq 0$ $x \in R$

Try this

4 Use the table above to draw a graph of the function

$f(x) = \dfrac{10}{x}$ $x \neq 0$ $-10 \leqslant x \leqslant 10$ $x \in R$
       The real numbers
       from −10 to 10 except 0
       including −10 and 10

Note: The graph of $f(x)$ is **symmetrical in the origin** (0, 0). That is, if a line is drawn from a point on the graph through the origin then it will meet another point on the graph at the same distance from the origin but in the opposite direction.

Let us see what this means by looking at the values of $f(x)$ at $x=2$ and $x=-2$.

$$f(2)=\frac{10}{2}=5 \qquad f(-2)=\frac{10}{-2}=-5$$

If we change the sign of a value of x (e.g. from 2 to -2), only the sign of $f(x)$ changes (from 5 to -5 in this example).

This function is said to be an **odd** function.

Try this

5 (a) Draw graphs of the following two functions on one set of axes

$$f(x)=x^2 \qquad -3\leqslant x\leqslant 3 \quad x\in R$$
$$g(x)=x^3 \qquad -3\leqslant x\leqslant 3 \quad x\in R$$

(b) Which one of the functions is an odd function?

(c) The other function is said to be an **even** function.
Complete the following statements about even functions:
The graph of an even function is symmetrical in the _____
i.e. if a point on the graph is reflected in the _____ *the image of the point is also on the graph*
If we change the sign of a value of x _____

Combination function

Let's look again at a graph of the function

$$f(x)=\frac{10}{x} \qquad x>0 \quad x\in R$$

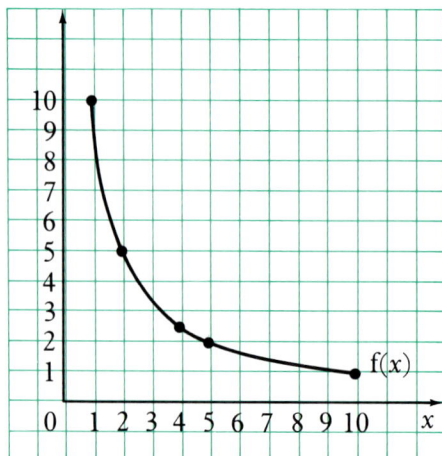

6

Now we draw a graph of the function
$g(x)=x \quad x\geqslant0 \quad x\in R$

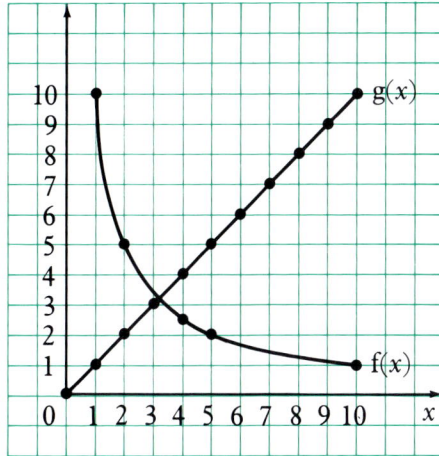

Finally, we 'add' points on the two graphs to get a graph of the 'combination' function

$h(x)=f(x)+g(x)$

$=\dfrac{10}{x}+x \quad x>0 \quad x\in R$

For example,
$h(1)=f(1)+g(1)$

$=\dfrac{10}{1}+1$

$=10+1=11$

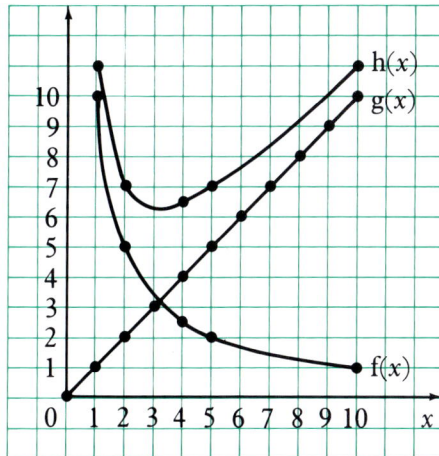

Try this

6 On one set of axes, draw graphs of the following functions:

$f(x)=\dfrac{10}{x} \quad x\neq0 \quad -10\leqslant x\leqslant10 \quad x\in R$

$g(x)=x \quad -10\leqslant x\leqslant10 \quad x\in R$

$h(x)=f(x)+g(x) \quad x\neq0 \quad -10\leqslant x\leqslant10 \quad x\in R$

The exponential function

The law of radioactive decay says that if M_0 is the mass of a radioactive material at the start of its decay then M_t, the mass which remains after time t, is given by

$M_t = M_0 e^{-kt}$

where k is a constant number whose value depends on the material. e is an important number in mathematics and we will refer to it later in the unit. We can see that M_t is a (rather unusual) function of t.

Let's look at a similar but simpler function

$f(n) = 2^n$ $n = 1, 2, 3, 4 \ldots$

(using n rather than x since we're dealing with integers).

A table of values

n	2^n	$(n, f(n))$
1	$2^1 = 2$	(1,2)
2	$2^2 = 4$	(2,4)
3	$2^3 = 8$	(3,8)
4	$2^4 = 16$	(4,16)
5	$2^5 = 32$	(5,32)

and a graph of $f(x) = 2^x$ $1 \leqslant x \leqslant 5$ $x \in R$

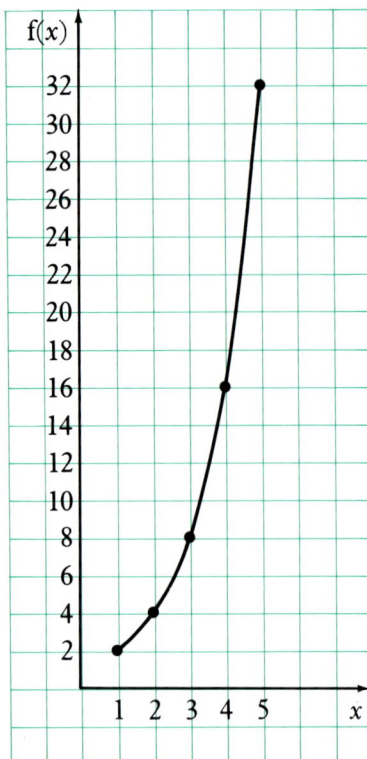

$f(x) = 2^x$ is an **exponential function**
x is the **exponent** and 2 is the **base**

Try this

6

7 (a) Make a table of values for the function
$g(n) = 3^n$ $n = 1, 2, 3, 4$

(b) Now draw a graph of the function
$g(x) = 3^x$ $1 \leqslant x \leqslant 4$ $x \in R$

Consider now the function
$f(n) = 2^n$ $n = 0, -1, -2, -3, -4 \ldots$

Reminders $a^0 = 1$ any number to the power 0 equals 1

$a^{-n} = \dfrac{1}{a^n}$ e.g. $2^{-3} = \dfrac{1}{2^3} = \dfrac{1}{8}$

A table of values...   ...and a graph

n	2^n	$(n, f(n))$
0	$2^0 = 1$	$(0, 1)$
-1	$2^{-1} = \frac{1}{2}$	$(-1, \frac{1}{2})$
-2	$2^{-2} = \frac{1}{4}$	$(-2, \frac{1}{4})$
-3	$2^{-3} = \frac{1}{8}$	$(-3, \frac{1}{8})$
-4	$2^{-4} = \frac{1}{16}$	$(-4, \frac{1}{16})$
-5	$2^{-5} = \frac{1}{32}$	$(-5, \frac{1}{32})$

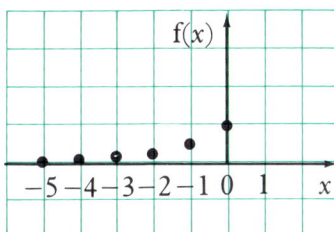

8 Draw a graph of the function

$f(x) = 2^x$ $\quad -5 \leqslant x \leqslant 5,\ x \in R$

9 Make a table of values and draw a graph of the function

$h(x) = 100 \times 3^{-x}$ $\quad 0 \leqslant x \leqslant 4,\ x \in R$

The function which we saw earlier, e^{-kt}, is an exponential function where the base, e, is called the **natural base**.

We won't go into this now, but you may have the chance to do so in the future if you take your study of mathematics further.

Inverse functions

Let's look at the function
$f(x) = 2x + 3 \quad x \in R$

A table of values...   ...and a graph

x	$2x + 3$	$(x, f(x))$
-3	-3	$(-3, -3)$
-2	-1	$(-2, -1)$
-1	1	$(-1, 1)$
0	3	$(0, 3)$
1	5	$(1, 5)$
2	7	$(2, 7)$
3	9	$(3, 9)$

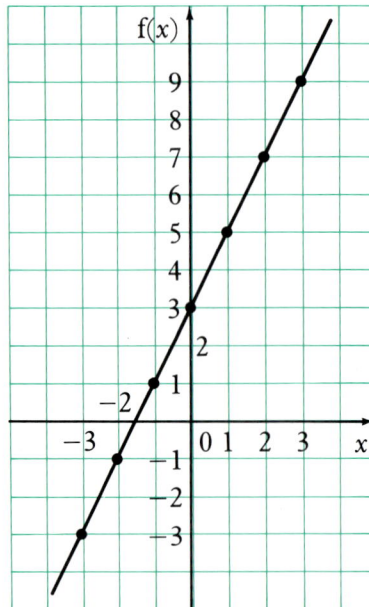

You will notice that each value of x goes to its own value of $f(x)$, different from any other.

We can say, therefore, that each value of $f(x)$ comes from **one and only one** value of x, and we should be able to find a function which relates $f(x)$ back to x.

Since we draw functions on an $x - y$ grid with $f(x)$ on the y-axis, let's include y in the function as follows:

$y = f(x) = 2x + 3 \quad x \in R$

6

$$x \xrightarrow[\text{multiply by 2}]{\times 2} 2x \xrightarrow[\text{add 3}]{+3} 2x+3$$

$$\frac{y-3}{2} \xleftarrow[\text{divide by 2}]{\div 2} y-3 \xleftarrow[\text{subtract 3}]{-3} y$$

This **inverse function** which relates y back to x is written as follows

$$x = f^{-1}(y) = \frac{y-3}{2} \qquad y \in R$$

Try these

10 Complete the missing parts ($\star$) in the following

$$y = f(x) = 5x - 4 \qquad x \in R$$

$$x \xrightarrow[\text{multiply by 5}]{\times 5} 5x \xrightarrow[\text{subtract 4}]{-4} 5x-4$$

$$\star \xleftarrow[\star]{\star} \quad \star \xleftarrow[\star]{\star} \quad y$$

$$x = f^{-1}(y) = \star \qquad y \in R$$

11 Find the inverse of each of the following functions

(a) $f(x) = 2x + 7 \qquad x \in R$

(b) $g(x) = \frac{1}{2}x - 6 \qquad x \in R$

(c) $h(x) = \frac{x+2}{3} \qquad x \in R$

Here is a table of values and a graph for the function
$f(x) = x^2 \qquad x \in R$

x	x^2	$(x, f(x))$
-3	9	$(-3, 9)$
-2	4	$(-2, 4)$
-1	1	$(-1, 1)$
0	0	$(0, 0)$
1	1	$(1, 1)$
2	4	$(2, 4)$
3	9	$(3, 9)$

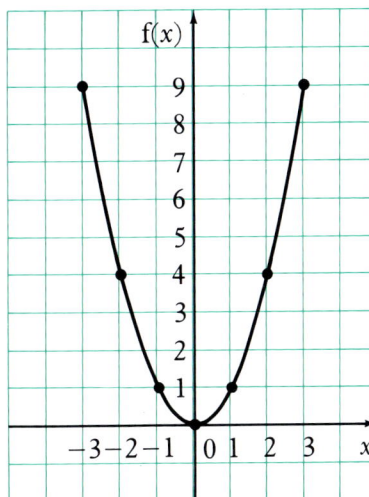

We cannot say, in this case, that each value of $f(x)$ comes from one and only one value of x.

In fact, each value of $f(x)$ is related to **two** values of x, e.g. $f(x)=4$ comes from $x=2$ or $x=-2$.

We cannot therefore write down an inverse function relating $f(x)$ back to x unless we restrict the function as follows:
$f(x)=x^2 \quad x \geqslant 0 \quad x \in R$

A graph of this function looks like this:

$$x \xrightarrow[\text{square}]{(\ )^2} x^2$$

$$\sqrt{y} \xleftarrow[\text{square root}]{\sqrt{\quad}} y$$

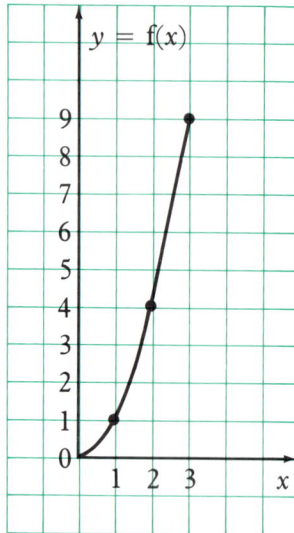

Since y exists only on the set of positive real numbers (and zero), the inverse of the function
$y=f(x)=x^2 \quad x \geqslant 0 \quad x \in R$
is
$x=f^{-1}(y)=\sqrt{y} \quad y \geqslant 0 \quad y \in R$

Try these

12 Sketch a graph of the function

$f(x)=x^2 \quad x \leqslant 0 \quad x \in R$

and write down its inverse.

13 Complete the missing parts ($\star$) in the following

$y=3x^2 \quad x \geqslant 0 \quad x \in R$

$$x \xrightarrow[\text{square}]{(\ )^2} x^2 \xrightarrow[\text{multiply by 3}]{\times 3} 3x^2$$

$$\star \xleftarrow[\star]{\star} \star \xleftarrow[\star]{\star} y$$

$x=f^{-1}(y)=\star \quad y \in R$

14 Find the inverse of each of the following functions

(a) $f(x)=\frac{1}{2}x^2 \quad x \geqslant 0 \quad x \in R$

(b) $g(x)=x^2-1 \quad x \geqslant 0 \quad x \in R$

(c) $h(x)=\frac{1}{2}x^2-1 \quad x \geqslant 0 \quad x \in R$

Now try this...

Earlier in this unit, we looked at the 'combination' function

$$h(x)=f(x)+g(x)$$
$$=\frac{10}{x}+x \qquad x\neq0 \quad x\in R$$

By drawing the three graphs on one set of axes, we see that the graph of $h(x)$ is contained 'between' the graphs of $f(x)$ and $g(x)$.

Investigate the combination function

$$p(x)=q(x)+r(x)$$
$$=\frac{10}{x}+(-x) \qquad x\neq0 \quad x\in R$$
$$=\frac{10}{x}-x$$

and comment on your findings.

Investigate also the combination functions

$$j(x)=\frac{10}{x}+x^2 \qquad x\neq0 \quad x\in R$$

and

$$k(x)=\frac{10}{x}-x^2 \qquad x\neq0 \quad x\in R$$

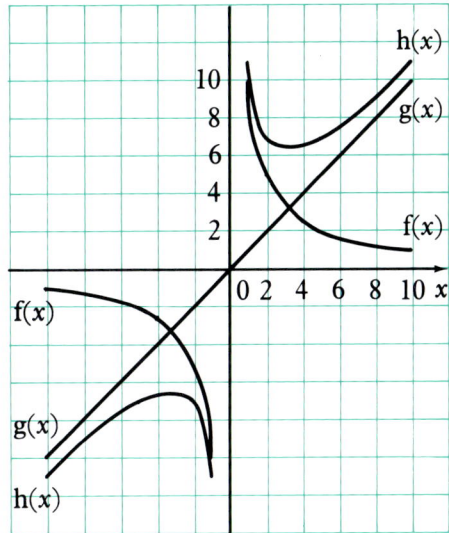

KEY QUESTIONS

K1 $\quad -2<x\leqslant3 \qquad x\in R$

can be shown in this way

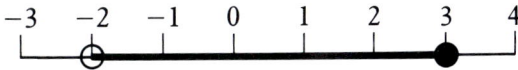

Show each of the following in a similar way.

(a) $-3\leqslant x<2 \qquad x\in R$

(b) $-1<x<1 \qquad x\in R$

(c) $-3\leqslant x\leqslant2 \qquad x\in R$

K2 For each of the following functions, calculate $f(1)$ and $f(-1)$ and state whether the function is odd, even or neither odd nor even.

(a) $f(x)=2x \qquad x\in R$

(b) $f(x)=(x+1)^2 \qquad x\in R$

(c) $f(x)=\dfrac{1}{x^4} \qquad x\in R$

K3 Find the inverse of each of the following functions.

(a) $g(x)=3x+4 \qquad x\in R$

(b) $h(x)=\dfrac{x-3}{2} \qquad x\in R$

(c) $j(x)=2x^3 \qquad x\in R$

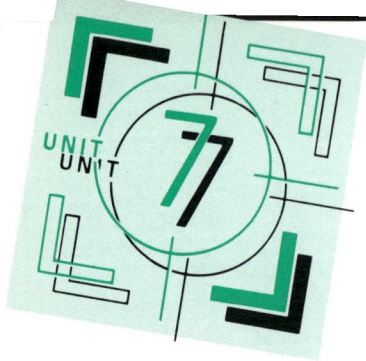

TRIGONOMETRIC FUNCTIONS AND GRAPHS

Here are some graphs of trigonometric functions.

$$y = \cos 4x \cdot \sin \tfrac{1}{4}x$$

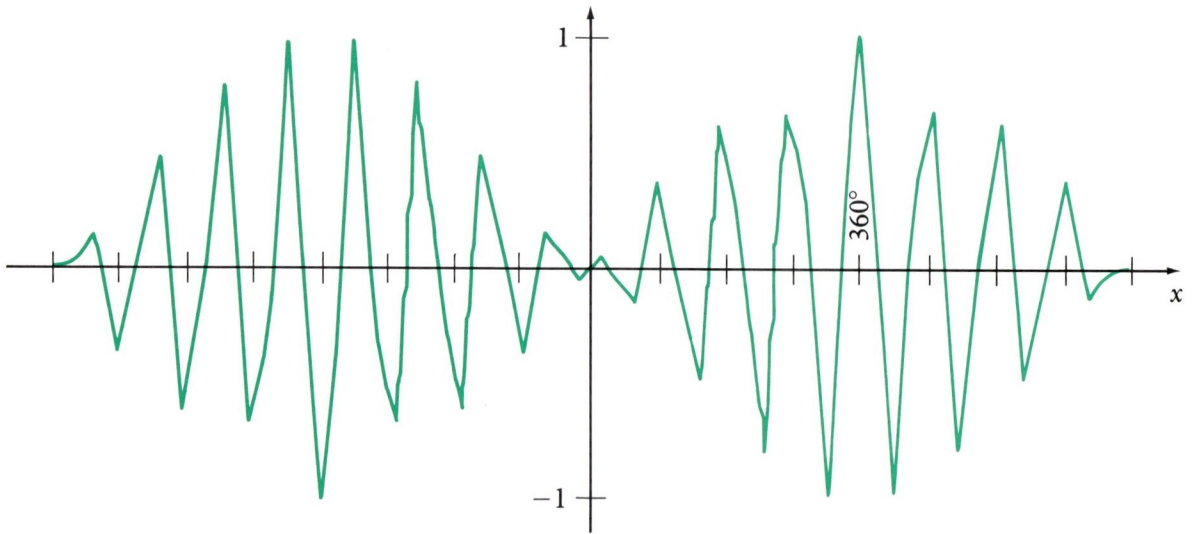

You may have seen such graphs on an **oscilloscope** in the science lab.

$$y = \sin^2 2x \cdot \cos^2 \tfrac{1}{4}x$$

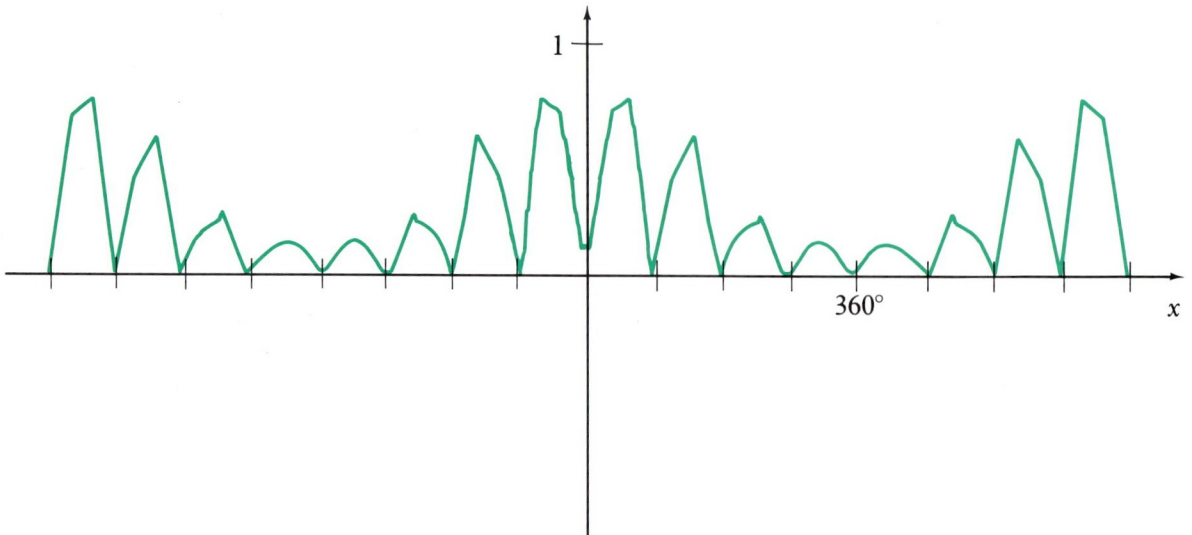

84

They are beyond the scope of this book......

. . . . . .but you should have learned enough by the end of this unit to understand how they come to look as they do.

$$y = \sin 4x \cdot \cos \tfrac{1}{4}x$$

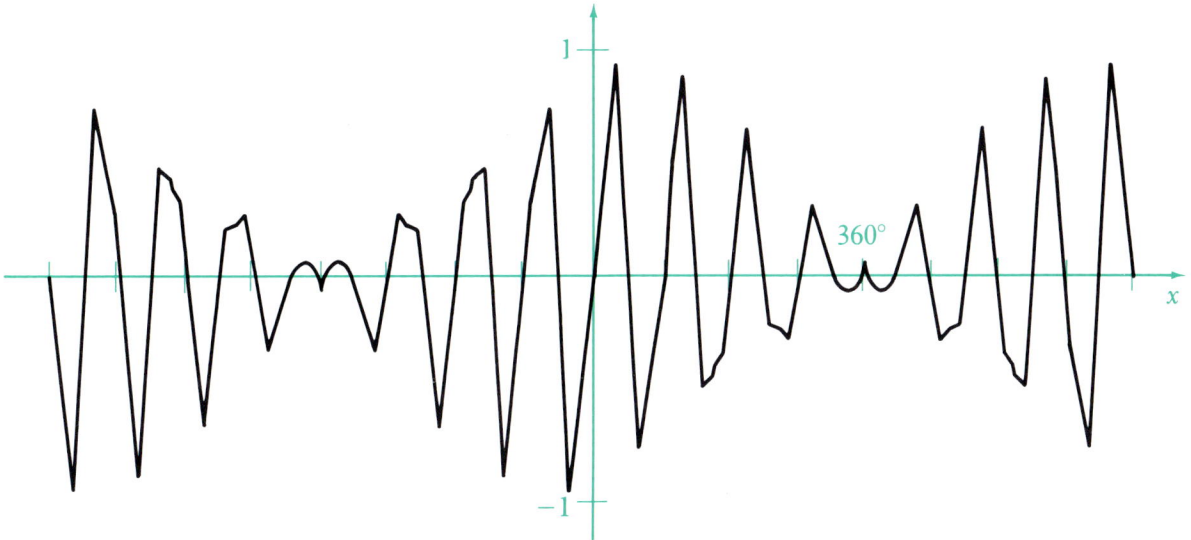

Many branches of science and engineering make use of graphs such as these.

$$y = -\cos^2 2x \cdot \sin^2 \tfrac{1}{4}x$$

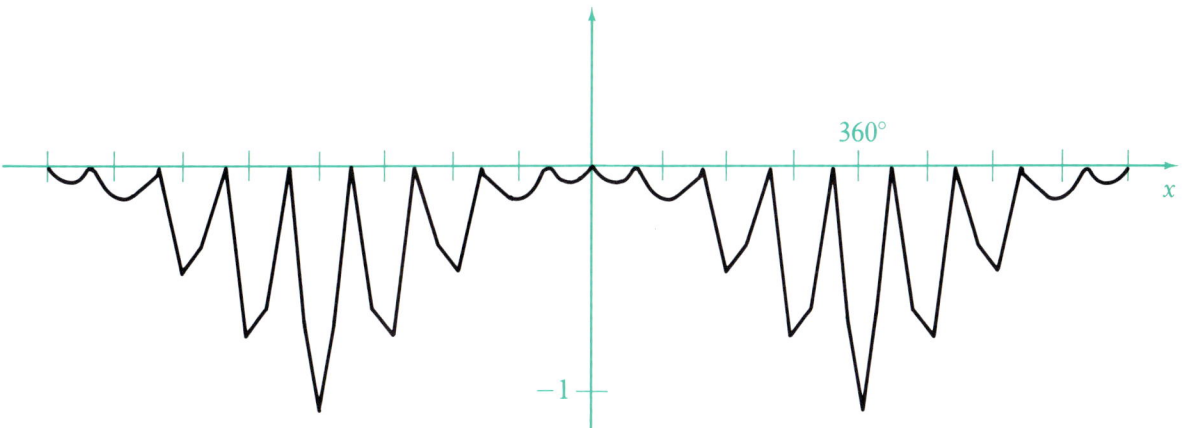

This unit introduces you to basic trigonometric functions and graphs. You may wish to progress to more complex cases in your later studies.

This disc is spinning at 1 revolution per second.
1 revolution = 360°
The disc is fixed at a point O on a horizontal line OX.
A point P has been marked on the circumference of the disc.

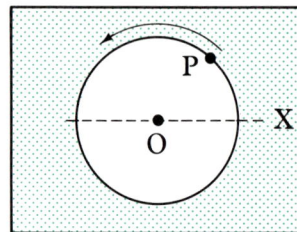

Some 'snapshots' are taken at the following times:

$\frac{1}{10}$ second $\frac{3}{10}$ second $\frac{5}{10}$ second $\frac{7}{10}$ second $\frac{9}{10}$ second

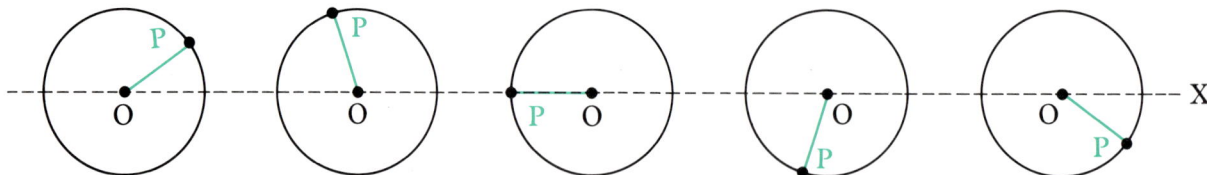

Try this

1   Draw 'snapshots' for the following times:

0 seconds, $\frac{2}{10}$ second, $\frac{4}{10}$ second,

$\frac{6}{10}$ second, $\frac{8}{10}$ second,   1 second

What is the height of P above the horizontal line OX in each of the snapshots?

Let's consider the snapshot taken at $\frac{1}{10}$th of a second.

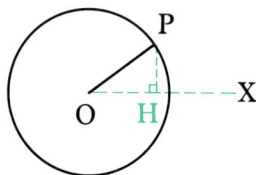

PH is the height of P above OX.
Triangle OHP is right-angled at H.

$$\frac{HP}{OP} = \sin POH$$

$HP = OP \times \sin POH$
(See *Yellow Book* for reminders if necessary)

Let us, for convenience, give the disc a radius of 1 unit,
and let PH = h and angle POH = A, so that
$HP = OP \times \sin POH$ becomes
  $h = 1 \times \sin A$
  $h = \sin A$

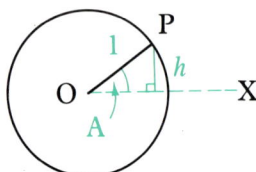

7

The height h is a function of the angle A and we may write
h(A) = sin A

A table of values of h for various values of A can be made by using a calculator with trigonometric functions.

Try this

2 Copy and complete this table.
The values of sin A are given to 2 significant figures.

A	sin A	(A,h(A))
0°	0	(0,0)
36°	0·59	(36°,0·59)
72°		
108°	0·95	(108°,0·95)
144°		
180°	0	(180°,0)
216°	−0·59	(216°,−0·59)
252°		
288°	−0·95	(288°,−0·95)
324°		
360°	0	(360°,0)

A graph of h(A) = sin A looks like this.

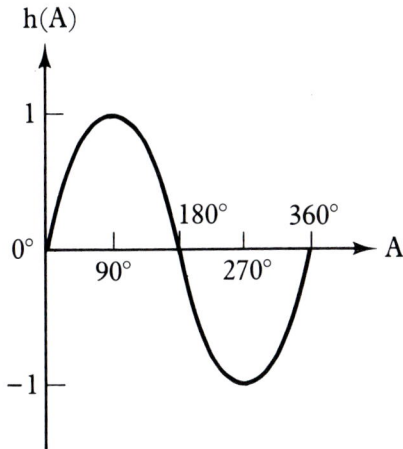

Perhaps you have already noted the following points from the table and graph.

The maximum value of sin A is 1; the minimum value of sin A is −1.

sin 0° = 0

sin 90° = 1   (from the graph)

sin 108° = sin 72°
If we look at the disc at 72° and at 108° we see that the height is the same, because in both cases the *acute* angle that OP makes with OX is 72°.

Similarly, **sin 144° = sin 36°**

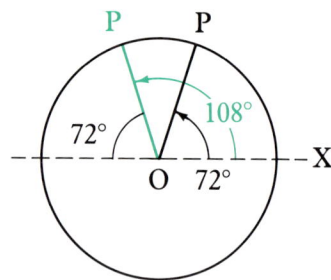

sin 180° = 0 (from the graph)

If A is between 0° and 180°, sin A is positive.

sin 216° is negative
sin 216° = −sin 36°
If we look at the disc at 36° and at 216° we see that the height is the same, because in both cases the *acute* angle that OP makes with OX is 36°, but at 216° P is *below* OX, making h *negative*.

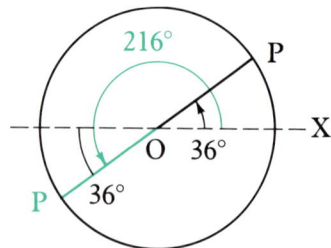

Similarly, **sin 252° = −sin 72°**

sin 270° = −1 (from the graph)

sin 288° is negative
sin 288° = −sin 72°
If we look at the disc at 72° and 288° we see that the height is the same, because in both cases the *acute* angle that OP makes with OX is 72°, but at 288° P is *below* OX, making h *negative*.

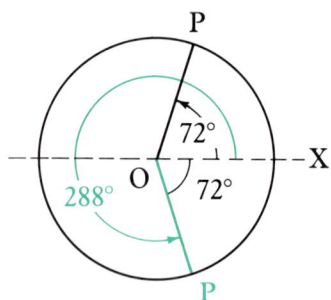

Similarly, **sin 324° = −sin 36°**

sin 360° = 0 (from the graph)

If A is between 180° and 360°, sin A is negative.

Let's summarise these results.

sin 0° = 0 ....0°		
sin 36° = 0·59 (to 2 significant figures)		
sin 72° = 0·95 (to 2 significant figures)	sin A POSITIVE	
sin 90° = 1 ....90°		
sin 108° = sin 72° (180° − 108° = 72°) sin 108° = 0·95		
sin 144° = sin 36° (180° − 144° = 36°) sin 144° = 0·59	sin A POSITIVE	
sin 180° = 0 ....180°		
sin 216 = −sin 36° (216° − 180° = 36°) sin 216° = −0·59		
sin 252 = −sin 72° (252° − 180° = 72°) sin 252° = −0·95	sin A NEGATIVE	
sin 270° = −1 ....270°		
sin 288° = −sin 72° (360° − 288° = 72°) sin 288° = −0·95		
sin 324° = −sin 36° (360° − 324° = 36°) sin 324° = −0·59	sin A NEGATIVE	
sin 360° = 0 ....360°		

Try this

3 (a) Copy and complete this table for an acute angle of 50°.

	$\sin 50° = ?$	(to 2 significant figures)
$\sin 130° = \sin ?°$	$\sin 130° = ?$	(to 2 significant figures)
$\sin 230° = -\sin ?°$	$\sin 230° = ?$	(to 2 significant figures)
$\sin 310° = -\sin ?°$	$\sin 310° = ?$	(to 2 significant figures)

(b) Complete a table similar to that in (a) for an acute angle of 70° giving your answers to 3 significant figures this time.

What about angles more than 360° and less than 0°?

After the disc has been spinning for $1\frac{1}{10}$ seconds, the angle is 396°.

From the diagram, we can see that the height of P above OX at 396° (360° + 36°) is the same as the height of P above OX at 36°.

Similarly, we can see that the height of P below OX at −36° (0° − 36°) is the same as the height of P below OX at 324° (360° − 36°).

We can check these results and others on a calculator.

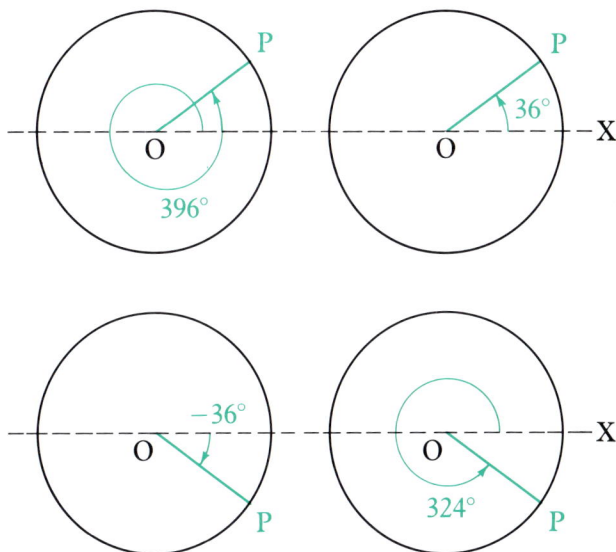

Try these

4 Copy and complete these tables.
The values of $\sin A$ are to be given to 2 significant figures.

(a)

A	396°	432°	468°	504°	540°	576°	612°	648°	684°	720°
$\sin A$										

(b)

A	−36°	−72°	−108°	−144°	−180°	−216°	−252°	−288°	−324°	−360°
$\sin A$										

5 Extend this graph of $\sin A$ to $720°$ and $-720°$.

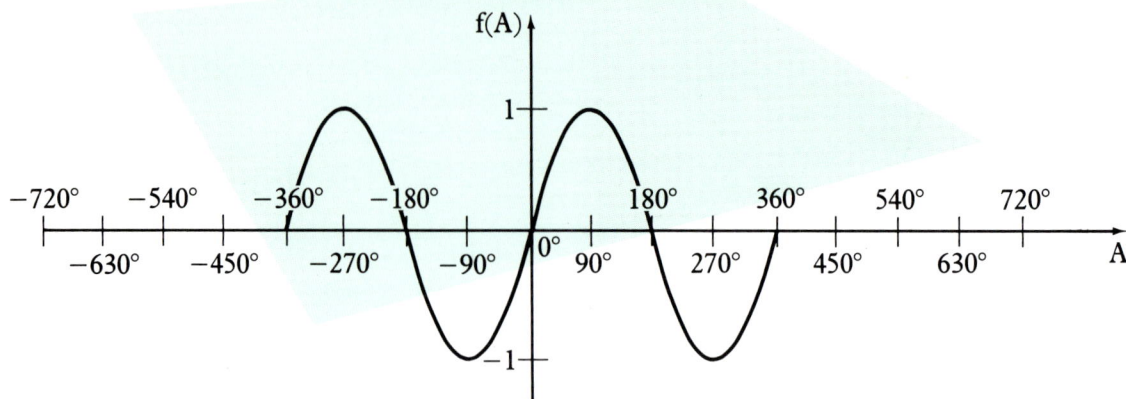

Note: $f(A) = \sin A$ is an **odd** function,
i.e. $f(-A) = -f(A)$, e.g. $\sin(-90°) = -1$, $\sin 90° = 1$.
The graph of $\sin A$ repeats itself every $360°$.
We say that $\sin A$ is **periodic** with a **period of $360°$**.

Let's now consider the length OH as the disc spins.

$$\frac{OH}{OP} = \cos A$$

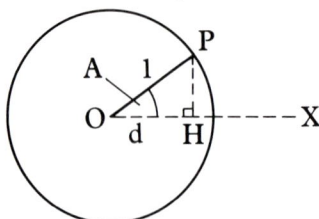

Let OH $= d$, then $d = \cos A$
d is a function of A and can be written
$d(A) = \cos A$

A table of values can be drawn up using a calculator.

Try this

6 Copy and complete this table.

A	$\cos A$	$(A, d(A))$
0°	1	(0,1)
36°	0·81	(36°,0·81)
72°		
108°	−0·31	(108°,−0·31)
144°		
180°	−1	(180°,−1)
216°	−0·81	(216°,−0·81)
252°		
288°	−0·31	(288°,0·31)
324°		
360°	1	(360°,1)

7

A graph of d(A) = cos A looks like this.

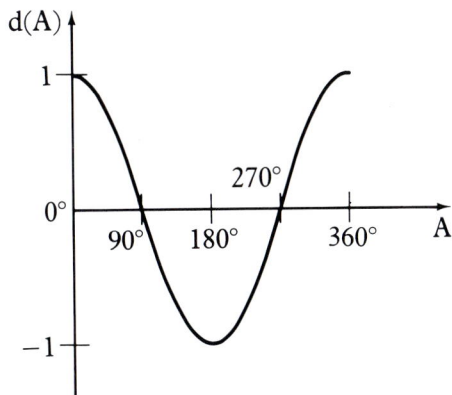

Notes

The maximum value of cos A is 1; the minimum value of cos A is −1.

cos 0° = 1

cos 90° = 0   (from the graph)

If A lies between 0° and 90°, cos A is positive.

cos 108° is negative
cos 108° = − cos 72°
If we look at the disc at 72° and 108° we see that d is the same, because in both cases the *acute* angle that OP makes with OX is 72°, but at 108° P is to the left of O, making d *negative*.

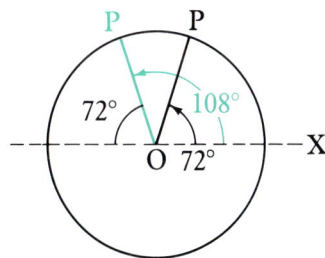

Similarly, **cos 144° = − cos 36°**

cos 180° = −1   (from the graph)

cos 216° is negative
cos 216° = − cos 36°
If we look at the disc at 36° and 216° we see that d is the same, because in both cases the *acute* angle that OP makes with OX is 36°, but at 216° P is to the left of O, making d *negative*.

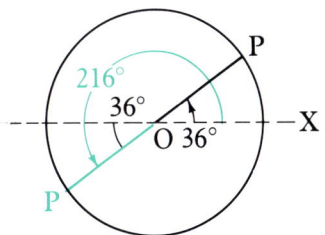

Similarly, **cos 252° = − cos 72°**

cos 270° = 0   (from the graph)

If A lies between 90° and 270°, cos A is negative.

cos 288° = cos 72°
If we look at the disc at 72° and 288° we see that d is the same, because in both cases the *acute* angle that OP makes with OX is 72°, and at 288° P is to the right of O, making d *positive*.

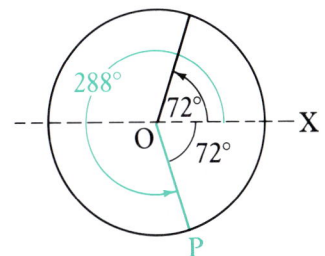

Similar, **cos 324° = cos 36°**

cos 360° = 1   (from the graph)

If A lies between 270° and 360°, cos A is positive.

Let's summarise these results.

$$\cos 0° = 1 \dots\dots\dots\dots\dots\dots\dots\dots\dots\dots\dots 0°$$
$\cos 36° = 0{\cdot}81$ (to 2 significant figures)
$\cos 72° = 0{\cdot}31$ (to 2 significant figures) $\cos A$ POSITIVE
$$\cos 90° = 0 \dots\dots\dots\dots\dots\dots\dots\dots\dots\dots\dots 90°$$
$\cos 108° = -\cos 72°$ $(180° - 108° = 72°)$ $\cos 108° = -0{\cdot}31$
$\cos 144° = -\cos 36°$ $(180° - 144° = 36°)$ $\cos 144° = -0{\cdot}81$ $\cos A$ NEGATIVE
$$\cos 180° = -1 \dots\dots\dots\dots\dots\dots\dots\dots\dots\dots 180°$$
$\cos 216° = -\cos 36°$ $(216° - 180° = 36°)$ $\cos 216° = -0{\cdot}81$
$\cos 252° = -\cos 72°$ $(252° - 180° = 72°)$ $\cos 252° = -0{\cdot}31$ $\cos A$ NEGATIVE
$$\cos 270° = 0 \dots\dots\dots\dots\dots\dots\dots\dots\dots\dots 270°$$
$\cos 288° = \cos 72°$ $(360° - 288° = 72°)$ $\cos 288° = 0{\cdot}31$
$\cos 324° = \cos 36°$ $(360° - 324° = 36°)$ $\cos 324° = 0{\cdot}81$ $\cos A$ POSITIVE
$$\cos 360° = 1 \dots\dots\dots\dots\dots\dots\dots\dots\dots\dots 360°$$

Try these

7 (a) Copy and complete this table for an acute angle of 70°.

	$\cos 70° = ?$	(to 2 significant figures)
$\cos 110° = -\cos ?°$	$\cos 110° = ?$	(to 2 significant figures)
$\cos 250° = -\cos ?°$	$\cos 250° = ?$	(to 2 significant figures)
$\cos 290° = \cos ?°$	$\cos 290° = ?$	(to 2 significant figures)

(b) Complete a table similar to that in (a) for an acute angle of 50° giving your answers to 3 significant figures this time.

8 Extend this graph of $\cos A$ to 720° and $-720°$.

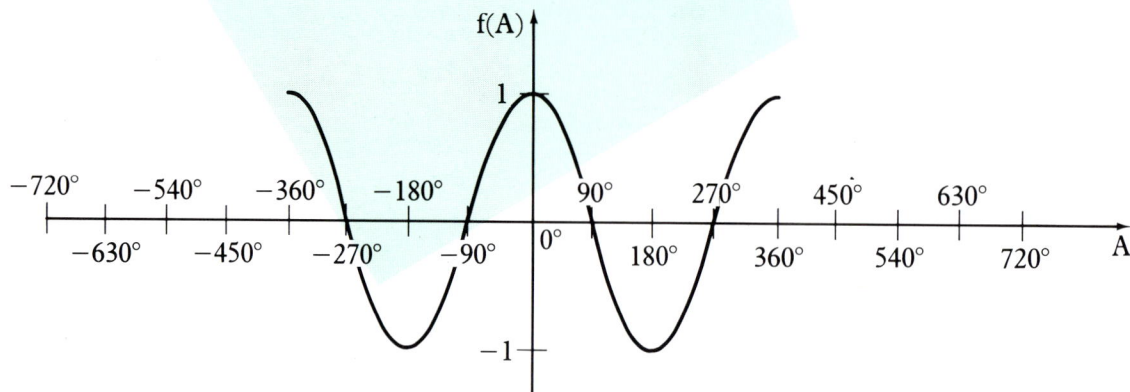

Note: $f(A) = \cos A$ is an **even** function,
i.e. $f(-A) = f(A)$, e.g. $\cos(-180°) = -1$, $\cos 180° = -1$.
$\cos A$ is **periodic** with a **period of 360°**.

Combination functions

7

Here, on one diagram, are sketch graphs of the functions

$s(x) = \sin x \qquad -720° \leqslant x \leqslant 720°$

$c(x) = \cos x \qquad -720° \leqslant x \leqslant 720°$

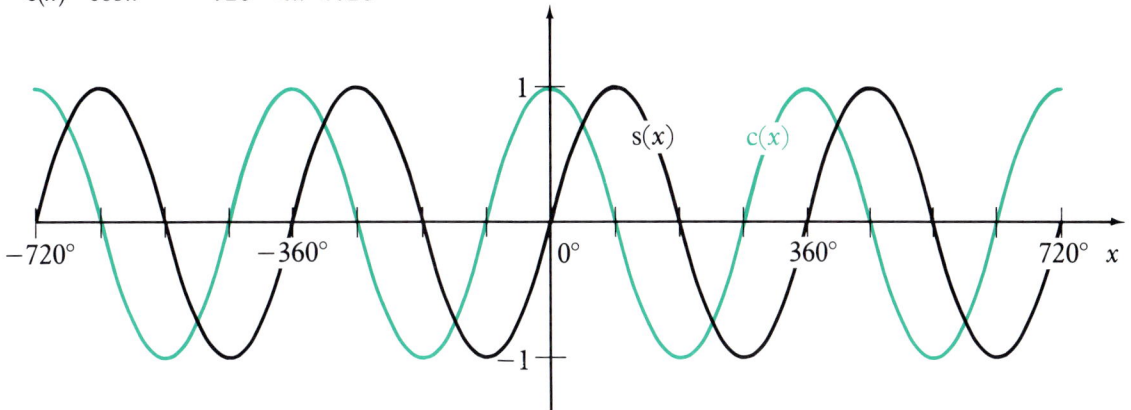

Consider now the task of sketching a graph of the combination function

$g(x) = \sin x + \cos x \qquad -720° \leqslant x \leqslant 720°$

Plotting some of the more obvious points gives the following graph.

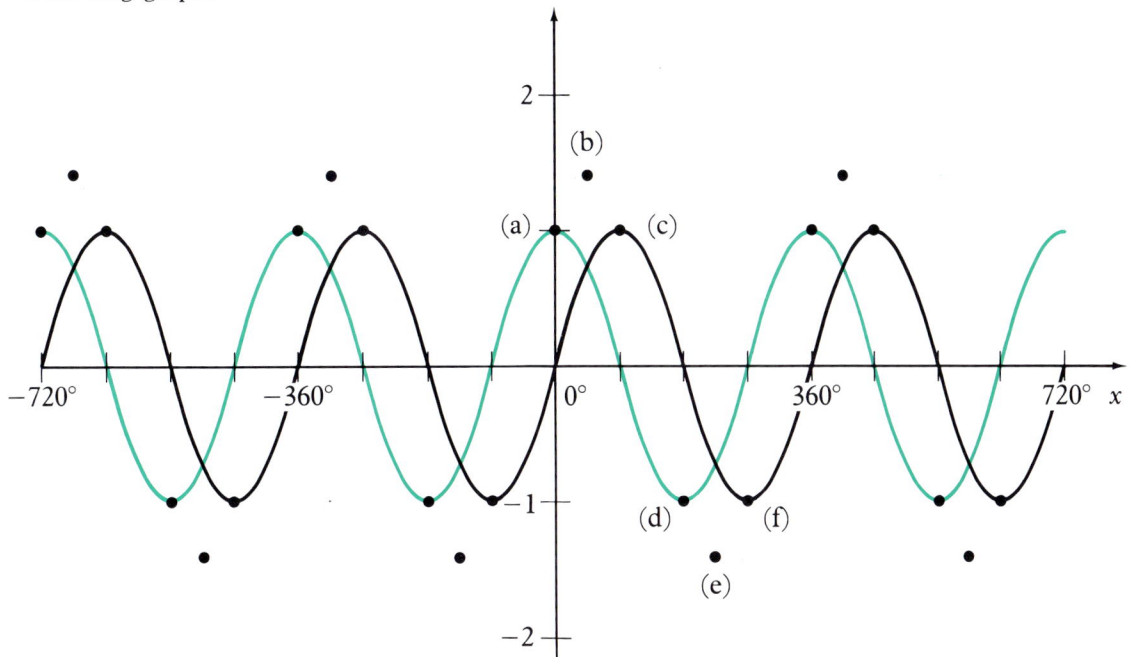

e.g. point (a): $\sin x = 0$, $\cos x = 1$, $\sin x + \cos x = 1$

point (b): $\sin x = Y$, $\cos x = Y$, $\sin x + \cos x = 2Y$ (twice the height)

point (c): $\sin x = 1$, $\cos x = 0$, $\sin x + \cos x = 1$

points (d), (e) and (f) as for (a), (b) and (c) but negative

Note that we have not calculated any values of $\sin x + \cos x$ but simply used the graphs of $\sin x$ and $\cos x$.

The completed graph of g(x) looks like this.

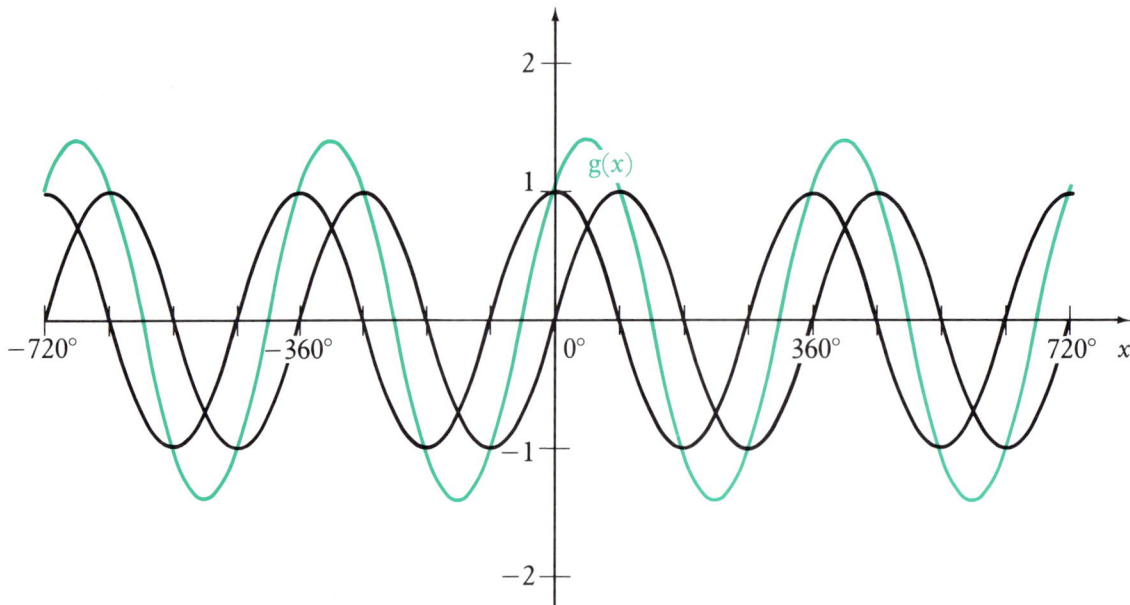

Note: in questions 9–14 you should sketch graphs for $-720° \leqslant x \leqslant 720°$.

Try this

9 (a) On one diagram, sketch graphs of $\sin x$ and $1 + \sin x$.
   (b) On one diagram, sketch graphs of $\cos x$ and $-\cos x$.
   (c) On one diagram, sketch graphs of $\cos x$ and $\cos x - 2$.
   (d) On one diagram, sketch graphs of $\sin x$, $\cos x$ and $\sin x - \cos x$.

Composite functions

7

Here, on one diagram, are sketch graphs of the functions

$s_1(x) = \sin x \qquad -720° \leqslant x \leqslant 720°$

$s_2(x) = 2\sin x \qquad -720° \leqslant x \leqslant 720°$

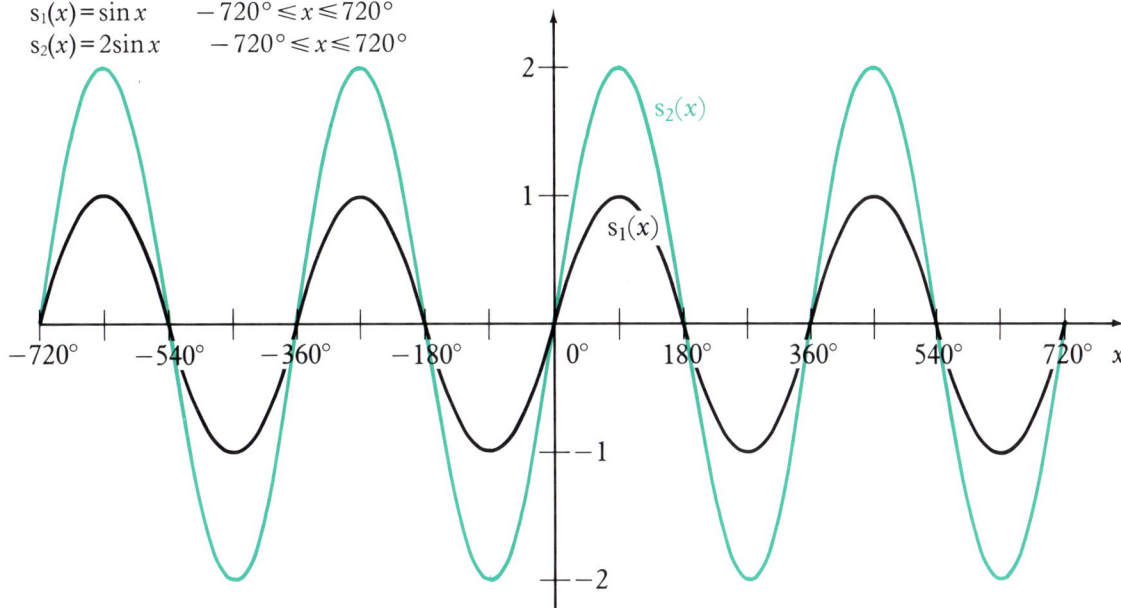

Note: (i) Both functions have a period of 360°.

(ii) The maximum value of $2\sin x$ is 2; the minimum value is -2.

This maximum value is called the **amplitude**.

The amplitude of $s_2(x)$ is 2.

The amplitude of $s_1(x)$ is 1.

Try this

10 (a) On one diagram, sketch graphs of $\cos x$ and $2\cos x$.

(b) On one diagram, sketch graphs of $\sin x$ and $\frac{1}{2}\sin x$.

Here, on one diagram, are sketch graphs of the functions

$c_a(x) = \cos x \qquad -720° \leqslant x \leqslant 720°$

$c_b(x) = \cos 2x \qquad -720° \leqslant x \leqslant 720°$

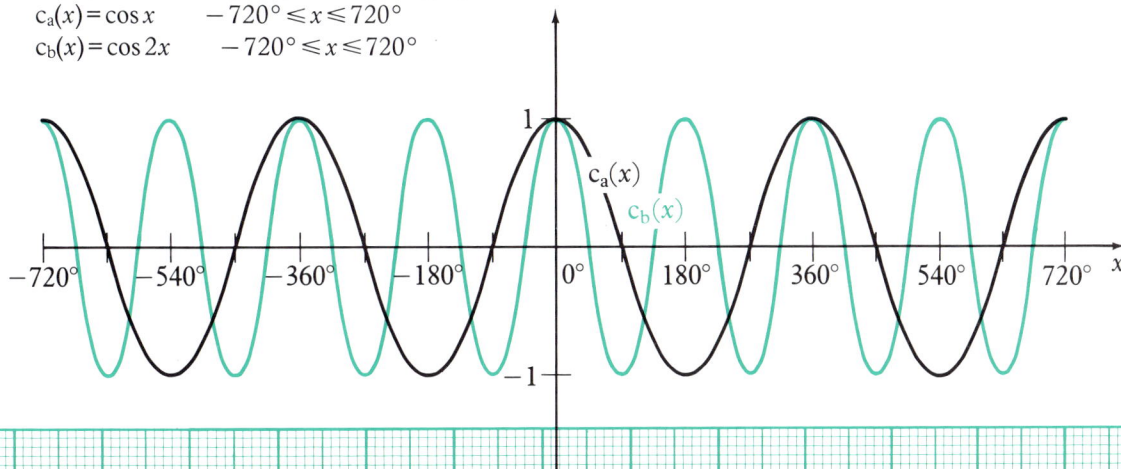

Note: $\cos x$ has a period of 360° and an amplitude of 1, as
we noted earlier.
$\cos 2x$ has a period of 180° and an amplitude of 1.

Try this

11 (a) On one diagram, sketch graphs of $\sin x$ and $\sin 2x$.
What is the period and amplitude of $\sin 2x$?

(b) On one diagram, sketch graphs of $\cos x$ and $\cos 3x$.
What is the period and amplitude of $\cos 3x$?

(c) On one diagram, sketch graphs of $\sin x$ and $\frac{1}{2}\sin 3x$.
What is the period and amplitude of $\frac{1}{2}\sin 3x$?

Here, on one diagram, are sketch graphs of the functions
$s_a(x) = \sin x \qquad -720° \leqslant x \leqslant 720°$
$s_b(x) = \sin\frac{1}{2}x \qquad -720° \leqslant x \leqslant 720°$

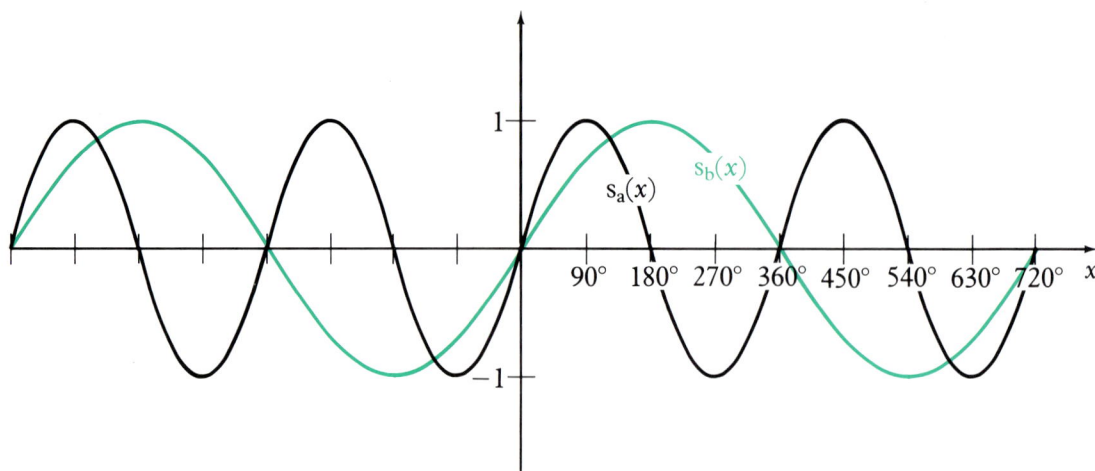

Note: $\sin x$ has a period of 360° and an amplitude of 1 as
we noted earlier.
$\sin\frac{1}{2}x$ has a period of 720° and an amplitude of 1.

Try this

12 (a) On one diagram, sketch graphs of $\cos x$ and $\cos\frac{1}{2}x$.
What is the period and amplitude of $\cos\frac{1}{2}x$?

(b) On one diagram, sketch graphs of $\sin x$ and $\sin\frac{1}{3}x$.
What is the period and amplitude of $\sin\frac{1}{3}x$?

(c) On one diagram, sketch graphs of $\cos x$ and $4\cos\frac{1}{3}x$.
What is the period and amplitude of $4\cos\frac{1}{3}x$?

Here, on one diagram, are sketch graphs of the functions
$c(x) = \cos x \qquad -720° \leqslant x \leqslant 720°$
$C(x) = \cos(x + 90°) \qquad -720° \leqslant x \leqslant 720°$

7

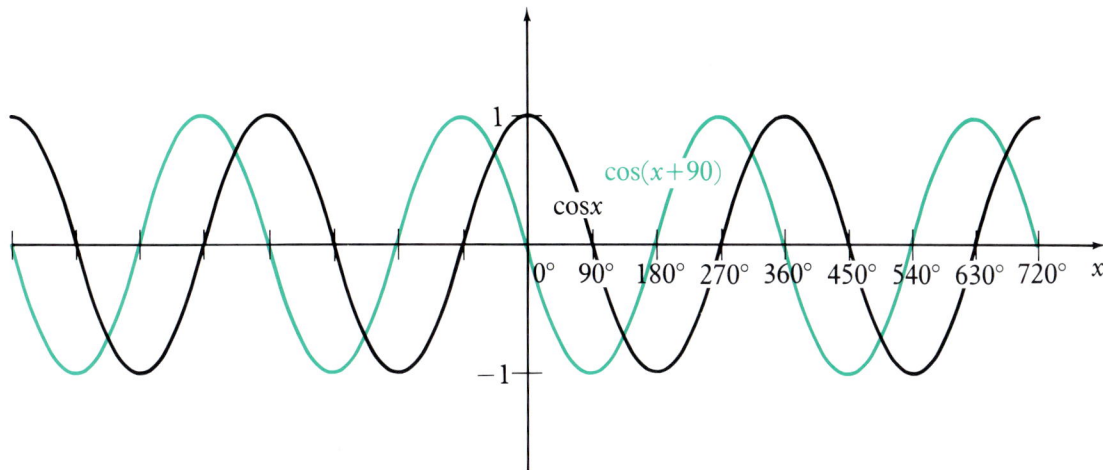

Note: Each graph has a period of 360° and an amplitude of 1.

The graph of $\cos(x + 90°)$ can be thought of as the graph of $\cos x$ 'shifted' 90° to the *left*. (Not to the right as you might have expected for +90°!)

Try this

13 (a) On one diagram, draw sketch graphs of $\sin x$ and $\sin(x + 90°)$.
(b) On one diagram, draw sketch graphs of $\cos x$ and $\cos(x - 90°)$.
(c) On one diagram, draw sketch graphs of $\sin x$ and $\sin(x + 180°)$.

Let's look at one more **composite** function
$f(x) = (\sin x)^2$

Again, without using a calculator, we can predict what the graph of $f(x)$ will look like.

$\sin 0° = 0 \qquad (\sin 0°)^2 = 0^2 = 0$
$\sin 90° = 1 \qquad (\sin 90°)^2 = 1^2 = 1$
$\sin 180° = 0 \qquad (\sin 180°)^2 = 0^2 = 0$
$\sin 270° = -1 \qquad (\sin 270°)^2 = (-1)^2 = 1$
$\sin 360° = 0 \qquad (\sin 360°)^2 = 0^2 = 0$

For some value of x between 0° and 90°, $\sin x = 0·5$.
For that value of x, $(\sin x)^2 = (0·5)^2 = 0·25$.

Note that, whatever the value of x, the value of $(\sin x)^2$ will be less than (or equal to) the value of $\sin x$.

Sketch graphs of $\sin x$ and $(\sin x)^2$ look like this.

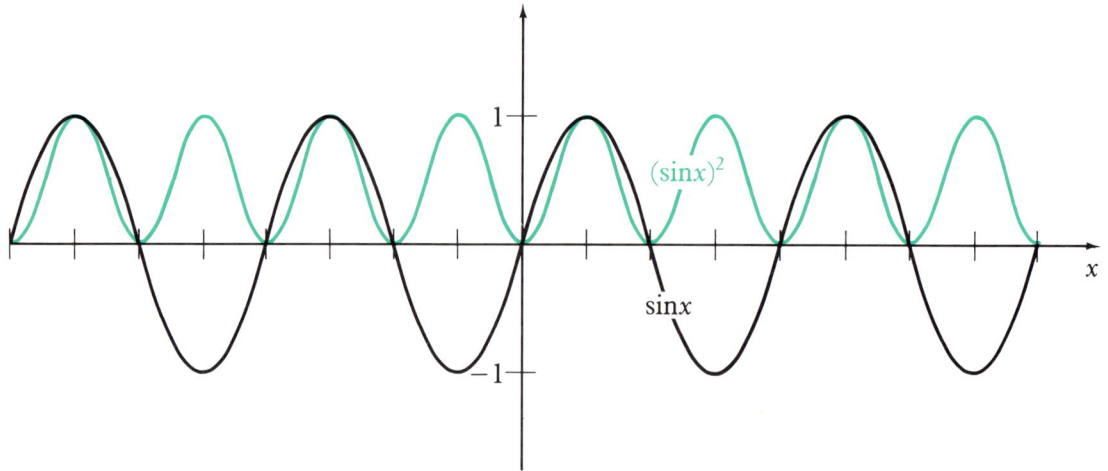

Note: (i) $(\sin x)^2$ is usually written as $\sin^2 x$ for convenience.
(ii) $\sin^2 x$ has a period of $180°$ and an amplitude of 1.

Try this

14 On one diagram, sketch graphs of $\cos x$ and $\cos^2 x$.
What is the period and amplitude of $\cos^2 x$?

Try this

15 Each of the following graphs represents a trigonometric function.
The function is written beside each graph in the form $a\sin(bx+c)$
or $a\cos(bx+c)$.
Write down the values of a, b, c as required for each.

(a) $a\sin bx$

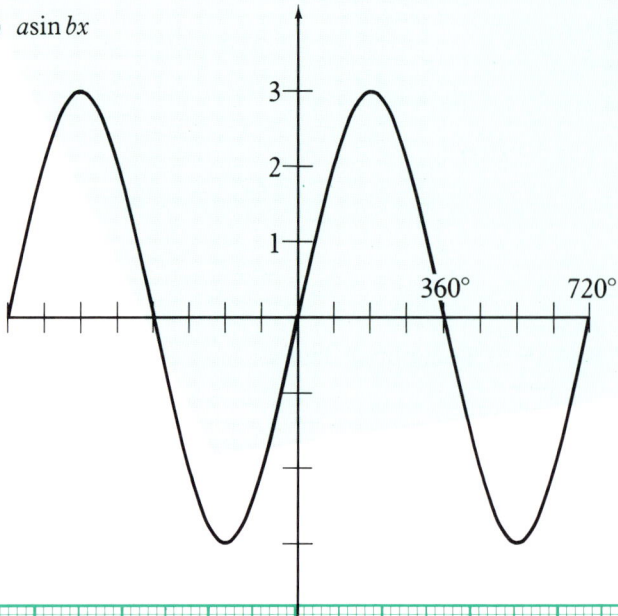

(b) $a\cos(x+c)$

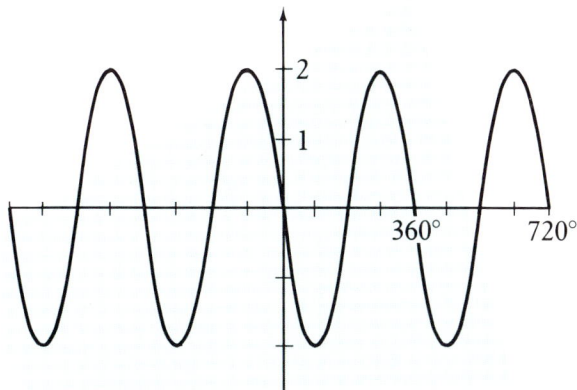

(c) $a\cos bx$

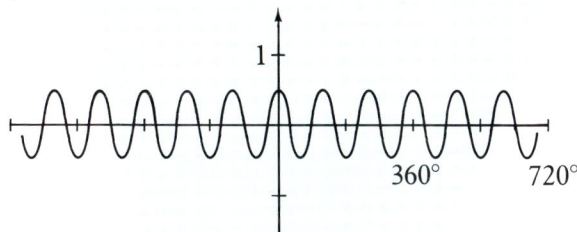

(d) $a\sin(x+c)$

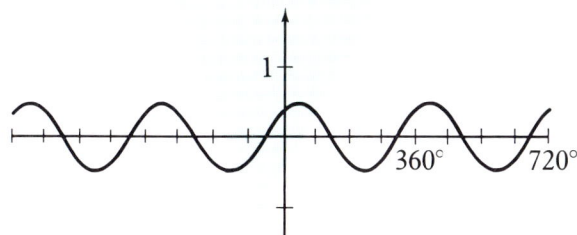

The tangent function

Try this

16 Copy and complete this table.
Use a calculator and give values of $\tan A$ to 2 significant figures.

A	0°	20°	40°	60°	80°	100°	120°	140°	160°	180°
$\tan A$	0	0·36			5·67	−5·67			−0·36	

A	200°	220°	240°	260°	280°	300°	320°	340°	360°
$\tan A$	0·36			5·67	−5·67			−0·36	

A sketch graph of t(A)=tan A looks like this.

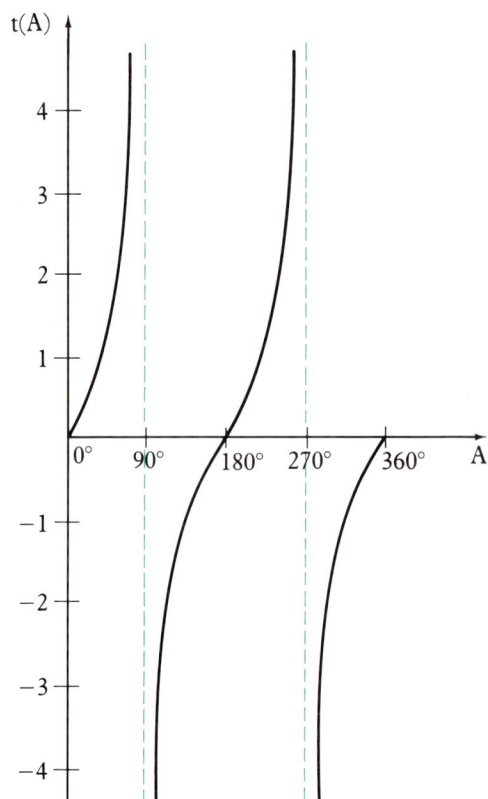

Note the following features of the graph of tan A.

tan 0° = 0

As A increases between 0° and 90°, tan A increases.

tan 89° = 57·3

When A = 90°, there is no value for tan A — your calculator gives you an 'error' message.

(We will look at this again in *Now Try This* at the end of the unit.)

tan 91° = −57·3

As A increases between 90° and 270°, tan A increases.

tan 180° = 0

tan 269° = 57·3

When A = 270°, there is no value for tan A — your caclulator gives you an 'error' message.

tan 271° = −57·3

As A increases between 270° and 360°, tan A increases.

tan 360° = 0

The graph of tan A is periodic with a period of 180°.

What is the relationship between the tangent of angles in
the intervals 0°–90°, 90°–180°, 180°–270°, 270°–360°?

$\tan 0° = 0$ . . . . 0°
$\tan 20° = 0·36$ (to 2 significant figures)
$\tan 80° = 5·67$ (to 2 significant figures) $\tan A$ POSITIVE
$\tan 90°$ is not defined . . . . 90°
$\tan 100° = -\tan 80°$ $(180° - 100° = 80°)$ $\tan 100° = -5·67$
$\tan 160° = -\tan 20°$ $(180° - 160° = 20°)$ $\tan 160° = -0·36$ $\tan A$ NEGATIVE
$\tan 180° = 0$ . . . . 180°
$\tan 200° = \tan 20°$ $(200° - 180° = 20°)$ $\tan 200° = 0·36$
$\tan 260° = \tan 80°$ $(260° - 180° = 80°)$ $\tan 260° = 5·67$ $\tan A$ POSITIVE
$\tan 270°$ is not defined . . . . 270°
$\tan 280° = -\tan 80°$ $(360° - 280° = 80°)$ $\tan 280° = -5·67$
$\tan 340° = -\tan 20°$ $(360° - 340° = 20°)$ $\tan 340° = -0·36$ $\tan A$ NEGATIVE
$\tan 360° = 0$ . . . . 360°

Try these

17 (a) Copy and complete this table for an acute angle of 30°

	$\tan 30° = ?$	(to 2 significant figures)
$\tan 150° = \tan ?°$	$\tan 150° = ?$	(to 2 significant figures)
$\tan 210° = \tan ?°$	$\tan 210° = ?$	(to 2 significant figures)
$\tan 330° = -\tan ?°$	$\tan 330° = ?$	(to 2 significant figures)

(b) Complete a table similar to that in (a) for an acute angle of
60° giving your answers to 3 significant figures this time.

18 Extend this graph of $\tan A$ to $-360°$.

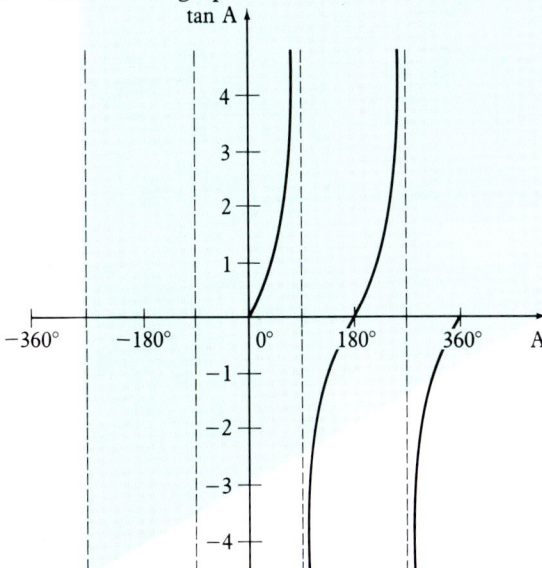

19 Do not use a calculator for this question.
Copy and complete the following table.

A	15°	54°	57°	123°	234°	345°
$\sin A$	0·26	0·81	0·84			
$\cos A$	0·97	0·59	0·54			
$\tan A$	0·27	1·38	1·54			

Now try this...

A The motion of a weight suspended by a spring can be described by the equation

$h = H\sin(kt)$

where

h is the height of the weight below the 'resting point',
t is the time for which the spring has been vibrating, and
H and k have fixed values which depend on the type of spring.

'resting point'

Let's consider a spring with H = 1 and k = 75.

When $t = 0$, $H\sin(kt) = 1 \times \sin(75 \times 0) = 1 \times \sin(0) = 1 \times 0 = 0$
When $t = 1$, $H\sin(kt) = 1 \times \sin(75 \times 1) = 1 \times \sin(75) = 1 \times 0{\cdot}97 = 0{\cdot}97$
When $t = 2$, $H\sin(kt) = 1 \times \sin(75 \times 2) = 1 \times \sin(150) = 1 \times 0{\cdot}50 = 0{\cdot}50$
When $t = 3$, $H\sin(kt) = 1 \times \sin(75 \times 3) = 1 \times \sin(225) = 1 \times -0{\cdot}71 = -0{\cdot}71$

Here are drawings of the spring at these times.

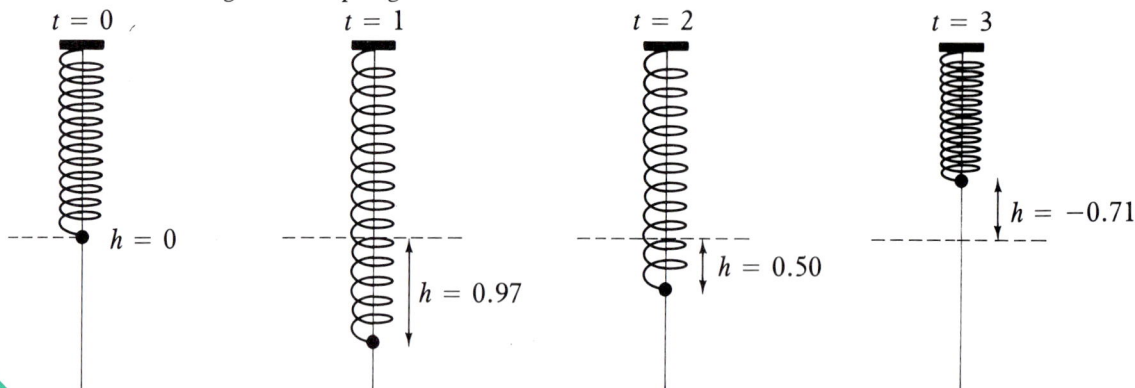

$t = 0$ $t = 1$ $t = 2$ $t = 3$

$h = 0$ $h = 0.97$ $h = 0.50$ $h = -0.71$

Make drawings of the spring at $t = 4$, and $t = 5$.

You should have noted that the spring takes between 4 and 5 seconds to make one complete **oscillation** — this time is called the **period**.

We are looking at a 'slow' spring here (assuming that t is measured in seconds).

What is the period of this spring?

Movement of objects such as springs and pendulums is known as **simple harmonic motion**.

Let's now consider a spring with k = 600 (and H = 1).

What is the period of this (k = 600) spring?

If those two springs start 'bouncing' together (at $t = 0$ and $h = 0$), how long will it be before they are together again at $h = 0$?

If two springs with periods of 2 seconds and 3 seconds start together, how long will it be before they are together again?

What is the relationship between the periods of two springs and the time it takes for them to be together again?

B We noted earlier in the unit that when we try to find tan 90° on the calculator we get an error message.

Why is this?

You may remember from the *Yellow Book* that in a right-angled triangle ABC

the tangent of angle $A = \dfrac{\text{the length of the side opposite angle A}}{\text{the length of the side adjacent to angle A}}$

$$\tan A = \frac{BC}{AB}$$

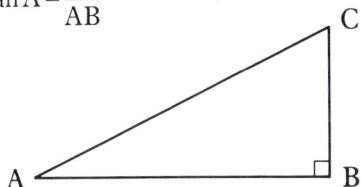

Use this definition of the tangent to answer the following questions.

1. What is the exact value of tan 45°?

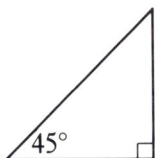

2. What happens to the value of tan A as A approaches 90°?

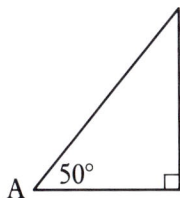

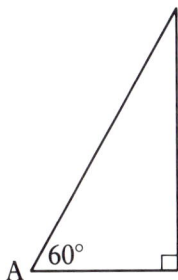

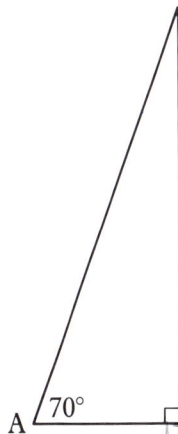

3. Why does tan 90° give an error message on the calculator?

KEY QUESTIONS

K1 Sketch a graph of $-\sin x$ $(-360° \leqslant x \leqslant 360°)$

K2 Sketch a graph of $\frac{1}{2}\cos x$ $(0° \leqslant x \leqslant 360°)$

K3 Sketch a graph of $\sin\frac{1}{4}x$ $(-720° \leqslant x \leqslant 720°)$

K4 Do not use a calculator for this question. Copy and complete the following table

A	25°	64°	87°	116°	267°	335°
$\sin A$	0·423	0·899	0·999			
$\cos A$	0·906	0·438	0·052			
$\tan A$	0·466	2·050	19·08			

K5 Each of the graphs in this question represents one of these functions.

$f_1(x) = \frac{1}{2}\cos 2x$ $-360° \leqslant x \leqslant 360°$
$f_2(x) = 2\cos\frac{1}{2}x$ $-360° \leqslant x \leqslant 360°$
$f_3(x) = \sin(x + 45)$ $-360° \leqslant x \leqslant 360°$
$f_4(x) = \sin(x - 45)$ $-360° \leqslant x \leqslant 360°$

Match each graph to its function.

A

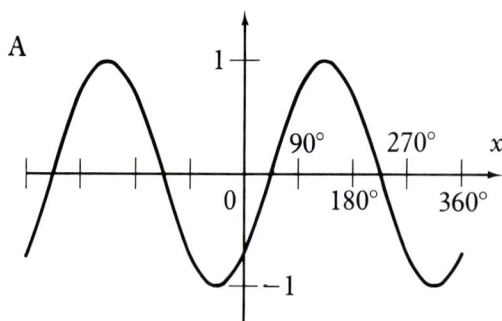

C

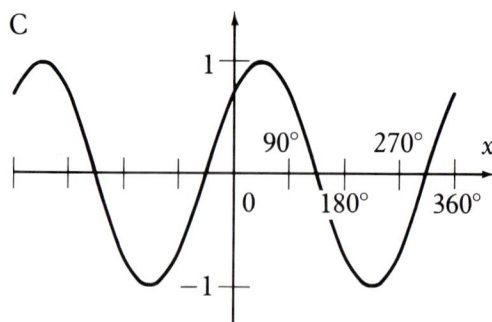

B

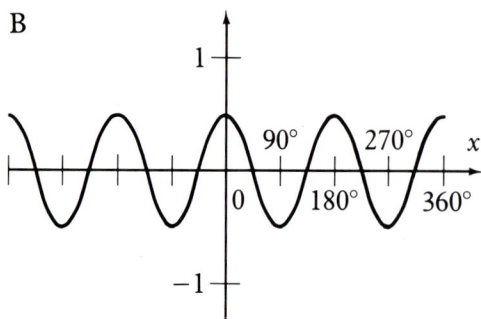

D

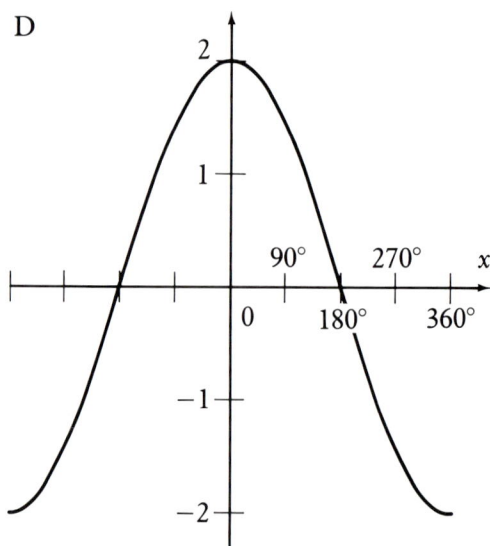

PROPORTIONALITY — FINDING THE GRAPH OR THE EQUATION

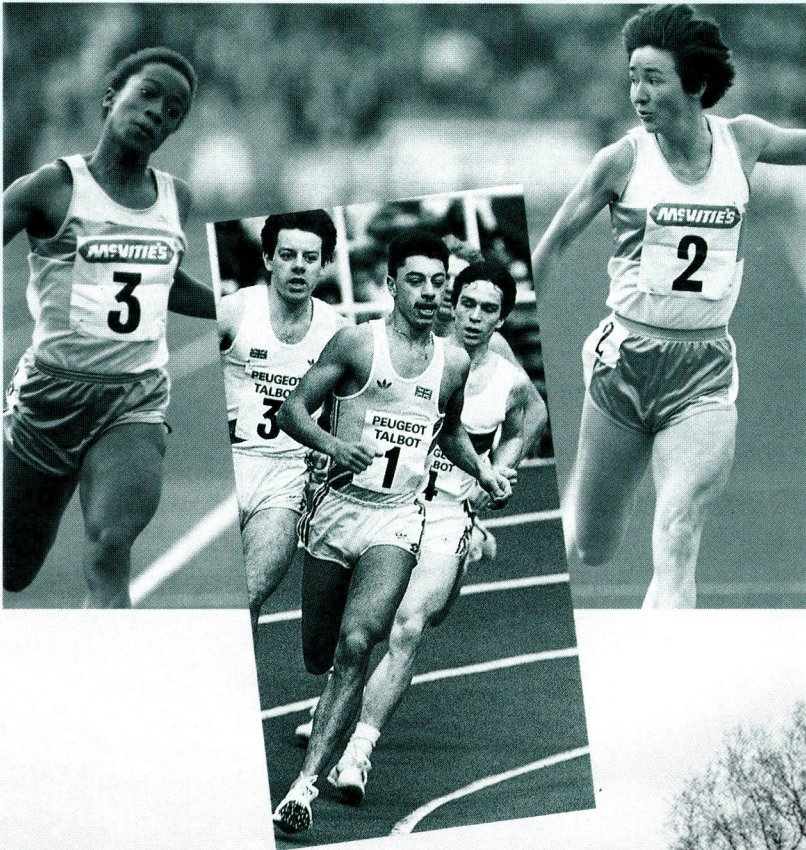

From the 100 m to the Marathon the world records for men are faster than those for women.

● But are the times related? Are they in proportion?

When the weather is overcast a photographer will select a number from the middle of this scale. This is an unusual set of numbers: 4, 5·6, 8, 11, 16, 22.

What effect do they have on the working of a camera?

By the end of this unit, you should have something to say about both of these questions.

Proportion — direct and inverse

In the *Yellow Book* there is a unit on proportion.
Let us revise two of the questions you met there.

A plumber starts to drain a 10 kilolitre tank full of water. After 5 minutes there is 8·5 kilolitres of water left in the tank.
How much longer will the tank take to drain?

In this example, the more time that passes the greater the amount of water which will have drained from the tank.

1·5 kilolitres have drained after 5 minutes.
8·5 kilolitres have still to drain.

1·5 kilolitres   5 mins

8·5 kilolitres   5 mins $\times \dfrac{8·5}{1·5}$

= 28 minutes to the nearest minute.

Let us look at the graphs of these examples.

Minutes	0	5	10	15	20	25	30
Kilolitres	0	1.5	3	4.5	6	7.5	9

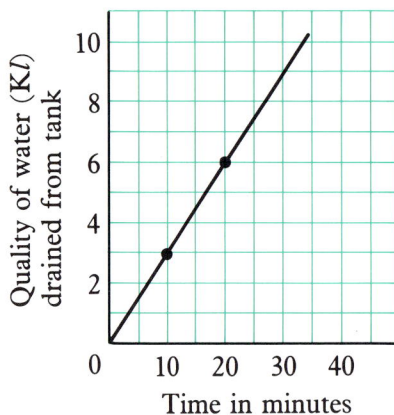

The amount of water drained is directly proportional to the time.

A contractor has been given the job of painting a school building, and has been told that it must be done in 4 days. He knows that the last time he arranged the same job, it took 12 painters 6 days to complete it.
How many painters should be put on the job this time?

In this example, the greater the number of painters employed, the less time it will take to complete the job.

To do the job in 6 days needed 12 painters.

To do the job in 1 day would need 12 × 6 = 72 painters.

To do the job in 4 days would need $\frac{72}{4}$ (= 18) painters.

Painters	1	2	4	6	8	9	12	18	36	72
Days	72	36	18	12	9	8	6	4	2	1

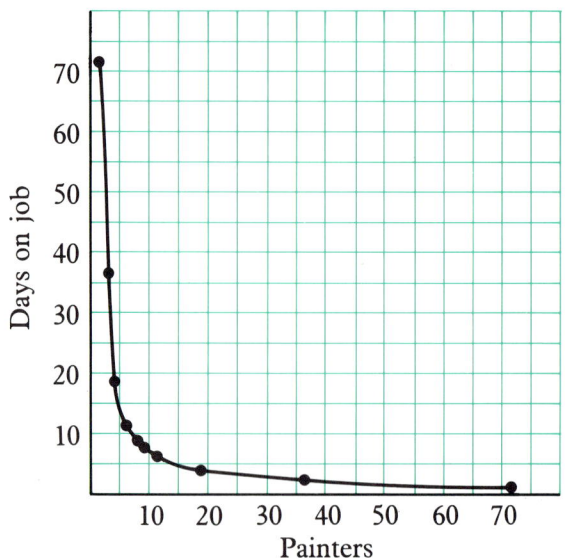

The number of days required to complete the job is inversely proportional to the number of painters.

The shorthand (symbol) for 'is proportional to' is $\propto$. So we can write

amount of water (w) $\propto$ time (t)

$w \propto t$

number of days (d) $\propto \dfrac{1}{\text{painters}(p)}$

$d \propto \dfrac{1}{p}$

(Note how we show an inverse. Remember that the inverse of 2 is $\frac{1}{2}$.)

The graph is a straight line which passes through the origin. You will remember from an earlier unit that equations of lines like this are of the form $y = mx$ where m is the gradient.

If the points were joined, your graph would be the same shape as a hyperbola. You met these in an earlier unit. Hyperbolas have equations of the form

$y = \dfrac{k}{x}$ where k is a constant.

This is called **DIRECT PROPORTIONALITY**.

This is called **INVERSE PROPORTIONALITY**.

Try these

1 In each of the following, state whether the two quantities are in direct proportion, in inverse proportion, or not in proportion:

(a) the number of people sharing a box of chocolates and the number of chocolates each can have, assuming a fair share for all;

(b) the profits of a school tuck shop and the number of packets of crisps sold;

(c) a person's age and his/her ability to play chess;

(d) the number of runners starting a Marathon and the number completing the course.

2 Using the symbol $\propto$ to represent 'is proportional to', write the following statements in mathematical shorthand:

(a) the gate receipts (R) at a football match and the number of spectators (S) attending;

(b) the time for a journey (t) and average speed (s);

(c) the amount of petrol (p) used and the length of a journey (l), assuming that the speed is constant.

3 (a) Match these equations to the graphs.

$$y = 2x - 3; \qquad y = \dfrac{2}{x} \qquad y = \tfrac{1}{2}x \qquad y = x^2 - 3$$

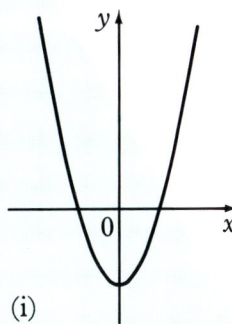

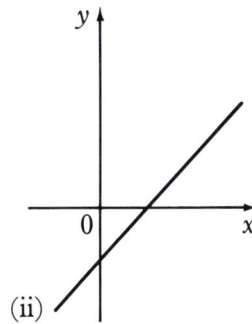

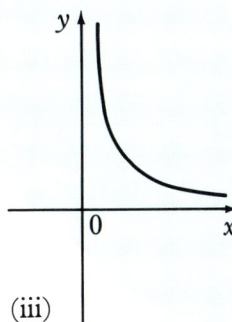

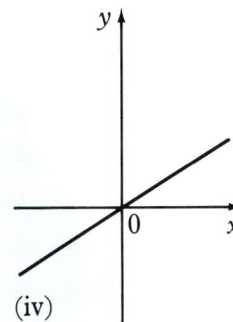

(i)

(ii)

(iii)

(iv)

(b) Which graph illustrates direct proportionality?

(c) Which graph illustrates inverse proportionality?

Direct proportionality (variation)

Let us look in some detail at direct proportionality. There is no new mathematics in this section but it brings together some pieces of work you have done earlier.

Let us look again at the table for draining water from a tank, which is an example of direct proportionality.

Kilolitres	1·5	3	4·5	6	7·5	9
Minutes	5	10	15	20	25	30

Note that the ratio of each number pair is a constant. (A constant is a member of the set of real numbers and includes negative numbers.)

$$\frac{\text{kilolitres}}{\text{Minutes}} \quad \frac{1·5}{5} \quad \frac{3}{10} \quad \frac{4·5}{15} \quad \frac{6}{20} \quad \frac{7·5}{25} \quad \frac{9}{30} \quad = \quad 0·3$$

So $\dfrac{\text{kilolitres}}{\text{minutes}} = 0·3$

The amounts (variables) are in direct proportion if the ratio is a constant.

Try this

4 Which of these tables of values show two variables which are in direct proportion?

(a)

y	2	8	20
x	1	4	10

(b)

p	1	2	4
q	16	8	4

(c)

b	6	−9	−3
a	−2	3	1

We have established earlier that the graph of two variables that are in direct proportion is a straight line through the origin.

EXAMPLE

A variable b is directly proportional to another variable a. The graph opposite shows the relationship between b and a.

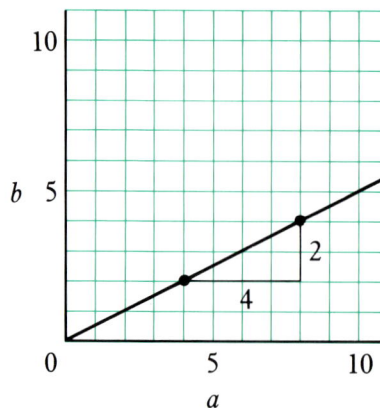

(4, 2) and (8, 4) are two of the points on this line. The ratio of $\dfrac{b}{a}$ is $\frac{2}{4} = \frac{4}{8} = 0·5$. The gradient of this line is also $\frac{2}{4} = 0·5$.

8

So if a variable b is directly proportional to another variable a, then the ratio $\frac{b}{a}$ is the same for every pair of corresponding values of a and b, and this ratio is equal to the gradient of the graph of (a, b).

For this example $\frac{b}{a} = 0 \cdot 5$.

This could also be written in the form:
$b = 0 \cdot 5a$

Most of the questions you will be asked about direct proportionality will involve forming an equation like the one above.

EXAMPLE

p is directly proportional to q. When $p = 18$, $q = 6$.

(a) Find an equation connecting p and q.

(b) Calculate p when $q = 4 \cdot 5$.

(c) Calculate q when $p = 7 \cdot 5$.

p is proportional to q, so $\frac{p}{q} = k$ where k is a constant.

This can be written as $p = kq$

(a) When $p = 18$, $q = 6$, so
$$18 = k \times 6$$
$$k = 3 \qquad \text{and}$$
$$p = 3q \quad \text{or} \quad \frac{p}{q} = 3$$

(b) When $q = 4 \cdot 5$, $p = 3q$, so
$$p = 3 \times 4 \cdot 5$$
$$p = 13 \cdot 5$$

(c) When $p = 7 \cdot 5$, $p = 3q$, so
$$7 \cdot 5 = 3q$$
$$q = 2 \cdot 5$$

Try these

5 q is directly proportional to p. When $q = 180$, $p = 9$.

(a) Find an equation connecting q and p.

(b) Calculate q when $p = 3 \cdot 2$.

(c) Calculate p when $q = 50$.

6 A travelling salesperson spends a good deal of time driving on motorways during which time he travels at a constant speed. So the cost of the petrol (p) is directly proportional to the distance travelled (d km). One day he travelled 360 km at a cost of £12·24.

(a) Find a formula connecting p and d.

(b) Calculate the cost of petrol for a distance of 220 km.

Arc of a circle

You will remember from an earlier unit that the formula for the circumference of a circle is:

$c = \pi d$

where d is the length of the diameter.

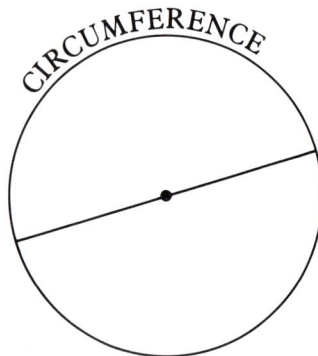

An arc of a circle is part of the circumference. For any two points on the circumference of a circle, there are two arcs — the major arc and the minor arc.

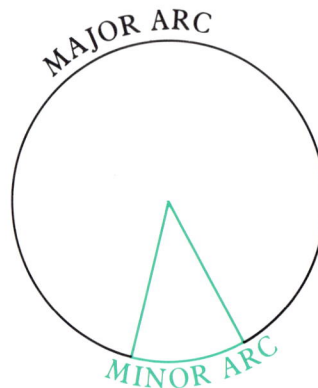

In the following questions assume that you are always calculating the length of the minor arc.

Given the diameter (or radius) of a circle, its circumference can be calculated. Let us see if our work on direct proportionality helps in calculating the length of an arc.

Try these

7 This circle has a diameter of 5 cm.

Calculate the length of the arc (minor) which subtends an angle of 90° at the centre of the circle. ($\pi = 3.14$)

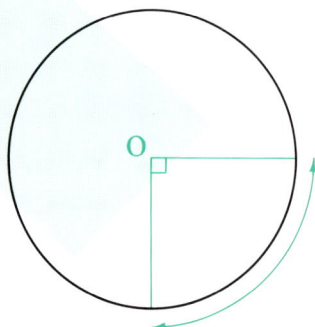

[writing]

8 This circle has a diameter of 4 cm.

(a) Calculate the length of the arc which subtends an angle of 55° at the centre of the circle.

(b) This calculation depends on two variables being in direct proportion. Which two variables?

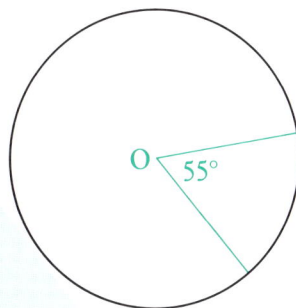

9 (a) Find a formula connecting the circumference (c) and the length of the arc of a circle (a). Let the angle at the centre be $b°$.

(b) Find a formula connecting the diameter (d) and the arc of a circle (a).

Assume that $\pi = 3.14$.

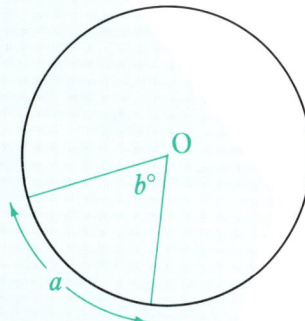

10 You are told the length of an arc and the size of an angle subtended at the centre. Is this sufficient information to calculate the circumference? Explain your answer.

11 Explain why the information shown here is sufficient to calculate the length of the arc.

This depends on two variables being in direct proportion. State the variables.

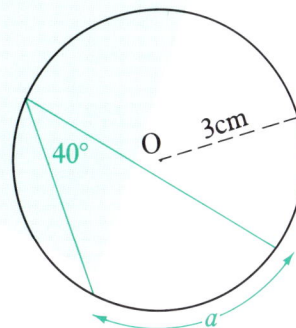

Other examples of direct proportionality

These cylinders have the same height (10 cm). In each case the radius is different.

The volume of a cylinder (V) is calculated by using the formula: $V = h\pi r^2$ where h is the height of the cylinder and r is the radius.

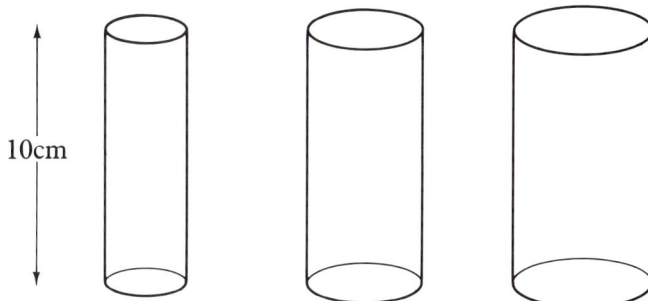

10cm

Let us look for evidence of proportionality between the volume of these cylinders and the radii. This table shows the volumes as the radius increases but the height remains constant at 10 cm.

radius (cm)	1	2	3	4	5
volume (cm³)	31·4	125·6	282·6	502·4	785

The ratio of $\dfrac{\text{volume}}{\text{radius}}$ is not constant.

$$\frac{31\cdot4}{1} \neq \frac{125\cdot6}{2} \neq \frac{282\cdot6}{3} \neq \frac{502\cdot4}{4} \neq \frac{785}{5}$$

The graph of volume against radius is certainly not a straight line.
In fact, the graph is a parabola.
So volume is not directly proportional to radius.

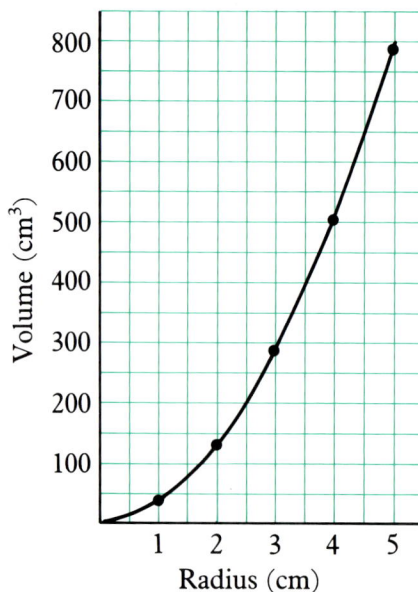

Let us examine the relationship between volume (V) and the radius squared (r^2).

radius²	1	4	9	16	25
volume	31·4	125·6	282·6	502·4	785

This time ratio of $\dfrac{\text{volume}}{\text{radius}^2}$ is constant.

$$\frac{31\cdot4}{1} = \frac{125\cdot6}{4} = \frac{282\cdot6}{9} = \frac{502\cdot4}{16} = \frac{785}{25} = 31\cdot4$$

8

The graph of volume against radius squared is a
straight line passing through the origin.

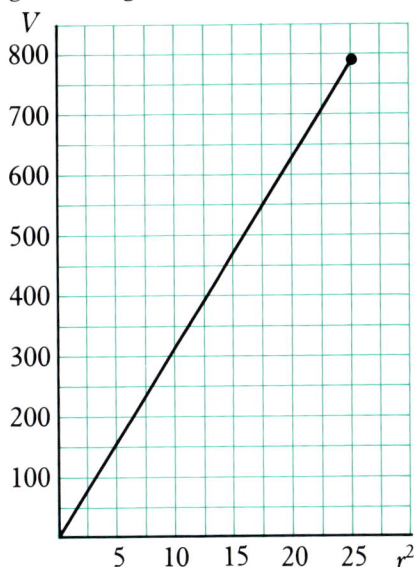

So, when cylinders have a constant height, the volume (V)
is directly proportional to the radius squared (r^2).

In symbols: $V \propto r^2$

The equation connecting V and r is $V = kr^2$.
Can you find the value of k?

Try these

12 Using the language and symbols of
proportionality, state the relationship
between the area (a) of a square and the
length (l) of its side.

13 State the relationship between the
volume (V) of a cube and the length (l)
of its side.

The speed of a cuckoo clock (speeding up or slowing down)
is determined by the length of the pendulum. The effective
length of pendulum is altered by moving the weight up and
down.

The time of the swing of a pendulum is the time taken for
the weight to go from one extreme to the other and back.

It has been proved by experiment that the time of a swing
is independent of the heaviness of the weight.

There is a connection between the length of a pendulum
and the time of the swing. This table was established by
experiment.

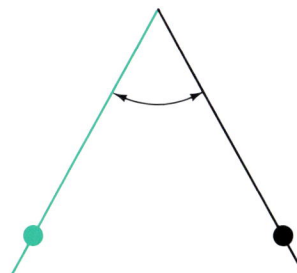

Time (t secs)	0·45	0·63	0·77	0·89	1
Length (l cm)	5	10	15	20	25

113

14 (a) Draw graphs of t against various powers of l (eg l^2, $l^{\frac{1}{2}}$, etc).

(b) State the relationship between line of swing and length of pendulum.

(c) Find the formula connecting t and l. Your answer will be approximate because the values in the table are from an experiment.

Area of a sector of a circle

The formula for the area of a circle is
$A = \pi r^2$ where r is the radius.

A segment is a 'slice' of the circle. It is bounded by two radii and an arc.
Any pair of radii makes two segments — major and minor.

In the following questions we shall concentrate on the minor segments.

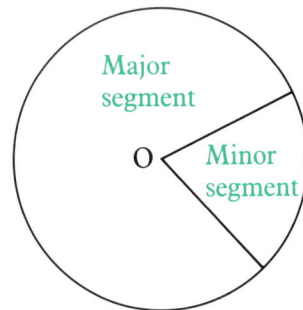

15 This circle has a radius of 3 cm.

Calculate the area of the segment that subtends an angle of 90° at the centre of the circle.

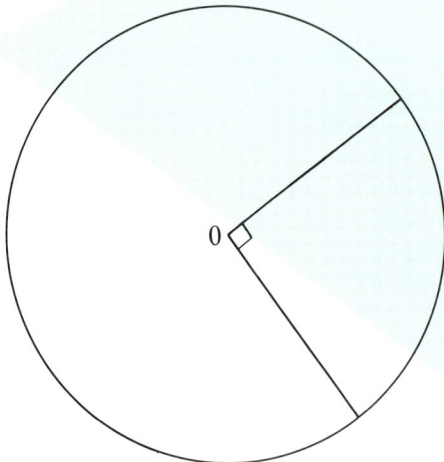

16 This circle has a diameter of 5 cm.

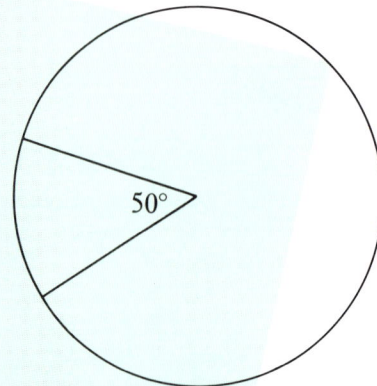

Calculate the area of the segment that subtends an angle of 50° at the centre of the circle.

17 Find a formula connecting the area of a circle (A) and the area of a segment (S).

Let the angle at the centre be $b°$.

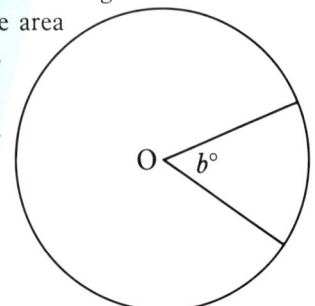

Inverse proportionality (variation)

8

Let us go back and take a look at inverse proportionality in some detail. Again, there is no new mathematics in this section.

Let us look again at the table for the number of painters and time taken to complete a job which we met at the start of the unit.

Painters	1	2	4	6	8	9	12	18	36	72
Days	72	36	18	12	9	8	6	4	2	1

Note that the product of each number pair is a constant,
i.e. $1 \times 72 = 2 \times 36 = 4 \times 18 = 6 \times 12 = 8 \times 9 = 9 \times 8 = 12 \times 6$
$= 18 \times 4 = 36 \times 2 = 72 \times 1 = 72$

Try this

18 Which of these tables show two variables which are in inverse proportion?

(a)

y	2	8	20
x	1	4	10

(b)

p	1	2	4
q	16	8	4

(c)

a	1	4	9	16
b	1	2	3	4

The graph, painters against number of days, is most certainly not a straight line.
This graph is called a hyperbola.

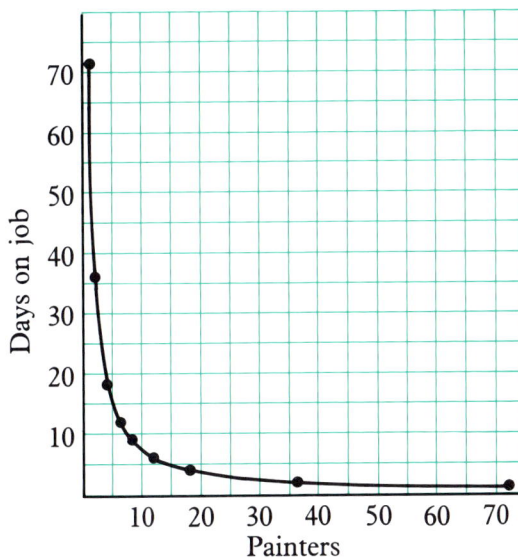

But the graph of days against the inverse of painters $\left(\dfrac{1}{\text{painters}}\right)$ is a straight line.

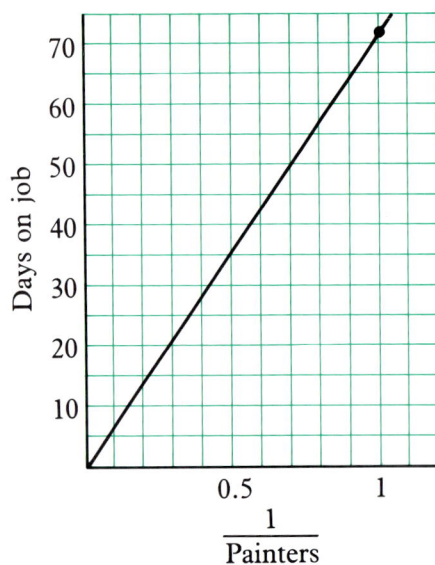

Days	72	36	18	12	9	8	6	4	2	1
$\dfrac{1}{\text{Painter}}$	$\frac{1}{1}$	$\frac{1}{2}$	$\frac{1}{4}$	$\frac{1}{6}$	$\frac{1}{8}$	$\frac{1}{9}$	$\frac{1}{12}$	$\frac{1}{18}$	$\frac{1}{36}$	$\frac{1}{72}$
$\dfrac{1}{\text{Painters}}$ (to 2 decimal places)	1	0·5	0·25	0·17	0·13	0·11	0·08	0·06	0·03	0·01

We could say that days d is proportional to the inverse of painters $\dfrac{1}{p}$ $\left(\text{i.e. } d \propto \dfrac{1}{p}\right)$;

or days is inversely proportional to number of painters $\left(\text{i.e. } d \propto \dfrac{1}{p}\right)$.

So if a variable d is inversely proportional to another variable p, then the product dp is the same for every pair of corresponding values of d and p.
For this example, $dp = 72$

This could also be written in the form $d = \dfrac{72}{p}$.

Most of the questions you will be asked about inverse proportionality will involve forming an equation like the one above.

EXAMPLE

a is inversely proportional to b. When $a = 4$, $b = 8$.
(a) Find an equation connecting a and b.
(b) Calculate a when $b = 2$.
(c) Calculate b when $a = 20$.

a is inversely proportional to b, so $ab=k$ (where k is a constant).

This can also be written as $a=\dfrac{k}{b}$

(a) When $a=4$, $b=8$, so
$$4\times 8=k$$
$$k=32$$

$$ab=32 \quad \text{or} \quad a=\dfrac{32}{b}$$

(b) When $b=2$, $\quad a\times 2=32$
$$a=16$$

(c) When $a=20$, $\quad 20\times b=32$
$$b=1\cdot 6$$

Try these

19 x is inversely proportional to y. When $x=7$, $y=4$.

(a) Find an equation connecting x and y.
(b) Calculate y when $x=2$.
(c) Calculate x when $y=10$.

20 p is inversely proportional to $\sqrt{q}$. When $p=2$, $q=16$.

(a) Find a formula connecting p and q.
(b) Calculate p when $q=25$.
(c) Calculate q when $p=4$.

21 m is inversely proportional to n^2. When $m=2$, $n=2$.

(a) Calculate m when $n=4$.
(b) Calculate n when $m=32$.

22 y is inversely proportional to x.

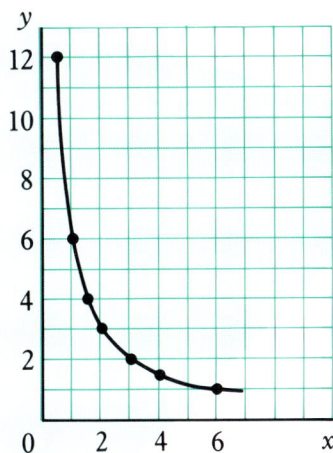

(a) Find an equation connecting y and x.

(b) Draw the graph of y against $\dfrac{1}{x}$.

23 A farmer's barn holds sufficient winter feeding to last a herd of 48 cattle for 52 days.

(a) If c is the number of cattle that can be fed for d days, find an equation connecting c and d.

(b) If the herd is increased by 30 cattle, how long will the same quantity of winter feed last.

Joint variation

Earlier we looked at ways of calculating the length of an arc of a circle. In question 9(b) you should have found the formula:

$$a = \frac{\pi bd}{360} \quad \text{or} \quad \frac{bd}{114 \cdot 6}$$

where a is length of arc, $b°$ the angle at centre, d the diameter, and $\pi = 3 \cdot 14$.

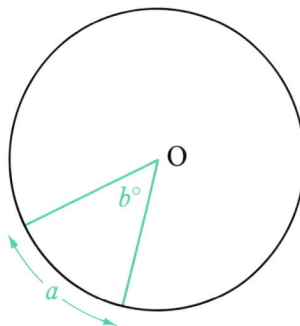

The length of the arc (a) depends on two variables: the size of the angle at the centre ($b°$) and the diameter of the circle (d).

As b increase, a increases (and vice versa).

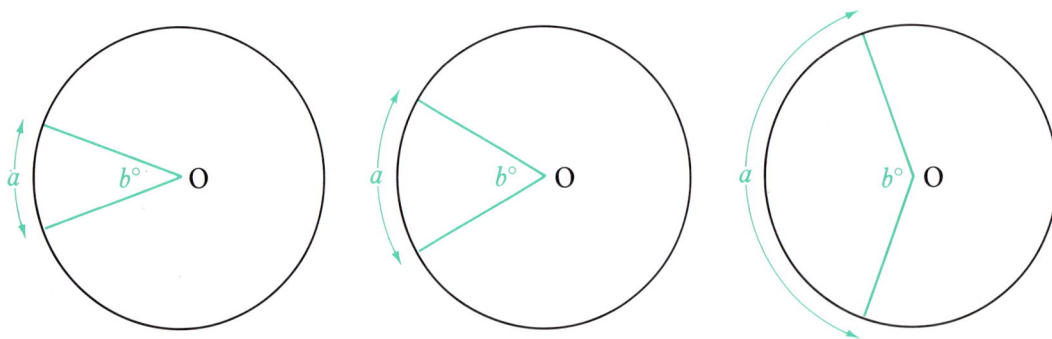

As d increases, a increases (and vice versa)

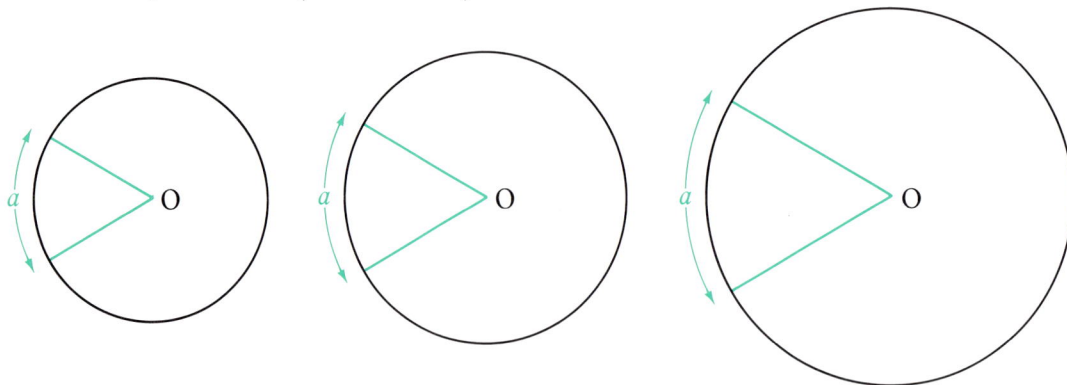

So, the length of an arc is directly (jointly) proportional to the angle subtended at the centre and the length of the diameter.

In symbols: $a \propto bd$

Equation: $a = \dfrac{bd}{114 \cdot 6}$

8

24 p is directly (jointly) proportional to q and r (i.e. $p \propto qr$). $p = 16$ when $q = 4$ and $r = 2$.

(a) Find the formula for p in terms of q and r.
(b) Calculate p when $q = 5$ and $r = 7$.
(c) Calculate q when $p = 25$ and $r = 2.5$.

25 A is directly proportional to l and b. $A = 21$ when $l = 3$ and $b = 1$.

(a) Find the formula for A in terms of l and b.
(b) Calculate A when $l = 1.7$ and $b = 3.2$.

26 x is directly proportional to y and the square of z. $x = 150$ when $y = 2$ and $z = 5$.

(a) Find the formula for x in terms of y and z.
(b) Calculate z when $x = 27$ and $y = 0.25$.

27 l is directly proportional to m and inversely proportional to n (i.e. $l \propto \dfrac{m}{n}$). $l = 10$ when $m = 20$ and $n = 8$.

(a) Find the formula for l in terms of m and n.
(b) Calculate l when $m = 16$ and $n = 8$.
(c) Calculate n when $l = 6$ and $m = 12$.

28 a is directly proportional to the square of b and inversely proportional to c. $a = 1$ when $b = 4$ and $c = 8$.

(a) Find the formula for a in terms of b and c.
(b) Calculate b when $a = 6$ and $c = 3$.

29 The wind resistance to a bus is directly proportional to the square of its speed. If the speed of the bus is increased from 20 mph to 60 mph, what happens to the wind resistance?

30 The weight (w) of a cylinder is directly proportional to the length (l) and the square of the radius (r).

(a) If the length is doubled, what happens to the weight?
(b) If the radius is doubled, what happens to the weight?

KEY QUESTIONS

K1 Indicate whether the table of figures is an example of direct or inverse proportionality.

(a)
p	24	2	6	4
q	1	12	4	6

(b)
a	10	15	7.5
b	4	6	2.5

K2 Indicate whether the graph is an example of direct or inverse proportionality.

(a)　(b)

K3 x is directly proportional to y. $x = 18$ when $y = 9$. Find y if $x = 9$.

K4 a is inversely proportional to b. $a = 7.5$ when $b = 4$. Find b if $a = 3$.

K5 p is directly proportional to the square of q and inversely proportional to r. $p = 3$ when $q = 6$ and $r = 4$.

(a) Find the formula for p in terms of q and r.
(b) Calculate r when $p = 1$ and $q = 3$.

Now try this...

A In the table below are world record times (1988) for running events from 100 m to the Marathon (approximately 40 km)

Distance	Men	Women
100 m	9·93 secs	10·76 secs
200 m	19·72	21·71
400 m	43·86	47·60
800 m	1:41·73	1:53·28
1000 m	2:12·18	2:30·6
1500 m	3:29·46	3:52·47
2000 m	4:50·81	5:28·69
3000 m	7:32·10	8:22·62
5000 m	12:58·39	14:37·33
10 000 m	27:13·81	30:13·74
Marathon	2:07:12·0	2:21:06·0

Can you find any relationship between the set of times given in the above table?

B The amount of light entering a camera is controlled by changing the f-stop setting.

The f-stop numbers are on the dial shown ⟶

The commonly used f-stop numbers are 4, 5·6, 8, 11 and 16.

$$\text{f-stop} = \frac{\text{focal length of lens } (l)}{\text{aperture diameter } (d)}$$

In a single-lens reflex camera the focal length is fixed. For the lens shown above, the focal length is 50 mm.

$$\text{So,} \quad \text{f-stop} = \frac{50}{\text{aperture diameter (mm)}}$$

Light entering the camera passes through the aperture (which is circular).

By considering the area of the aperture, examine the effect of different f-stop settings.

Can you suggest why the sequence of f-stop numbers is 4, 5·6, 8, 11 and 16?

TRIGONOMETRY

	PATH	
2		
		1

---

dense forest

Distance from 1 to 2	:	1·2km
Bearing of 2 from 1	:	290°

Orienteering is great fun. Most orienteering contests take place in forests. Competitors have to move from control point to control point (perhaps 20 in total) in the shortest possible time. It is not always advisable to go directly from one control point to the next. In a dense forest it is difficult to run at speeds in excess of 9 km/hr. Whereas in open ground (e.g. a path) a fit orienteer might average a speed of 14 km/hr.

● Look at the map above. Should an orienteerer run directly from control point 1 to control point 2? Would it be faster to run part of the way on the path?

By the end of this unit you should be able to help this orienteerer.

121

SOH CAH TOA

You will remember the 'word' SOH CAH TOA from the trigonometry unit in the *Yellow Book*.

It helps us to remember the meanings of the three trig ratios:

SOH $\left(\text{Sine} = \dfrac{\text{Opposite}}{\text{Hypotenuse}}\right)$ CAH $\left(\text{Cosine} = \dfrac{\text{Adjacent}}{\text{Hypotenuse}}\right)$ TOA $\left(\text{Tangent} = \dfrac{\text{Opposite}}{\text{Adjacent}}\right)$

Remember how we label the sides.

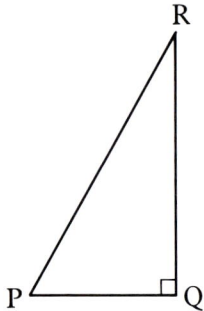 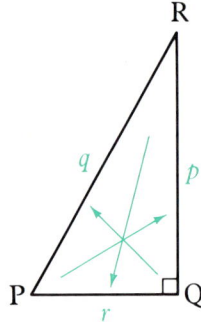

$\sin P = \dfrac{p}{q}$ $\cos P = \dfrac{r}{q}$ $\tan P = \dfrac{p}{r}$

Let us practice the use of the trig ratios.

In this unit, all answers should be given to 2 decimal places unless otherwise stated.

Try these

1   In each case find the length of side x (to two decimal places).

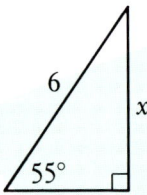   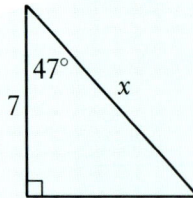   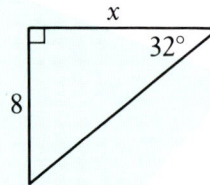   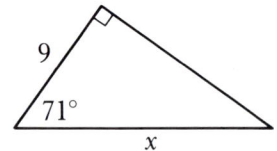

    (a)   (b)   (c)   (d)

2   In each case, find the size of angle x (to two decimal places).

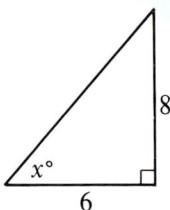   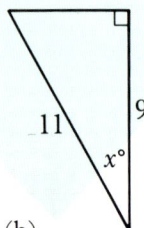   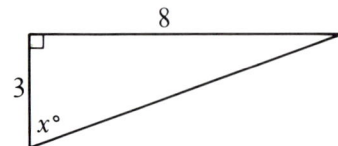

    (a)   (b)   (c)

3 In each case, calculate the length of all sides and sizes of angles (apart from those given).

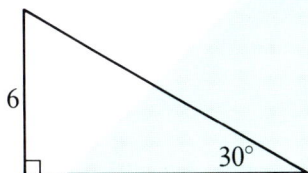

(a)

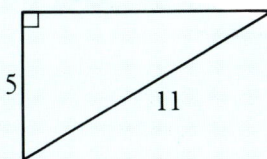

(b)

4 Before using a trig ratio, what is the minimum information which must be known in any triangle?

So far, all the triangles have contained a right angle. Let us try using the trig ratios in triangles which do not have a right angle.

EXAMPLE

Calculate the length of BC.

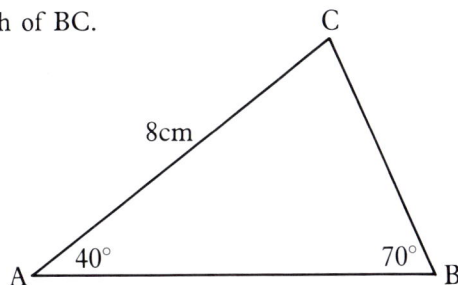

Step 1 Form two right-angled triangles.

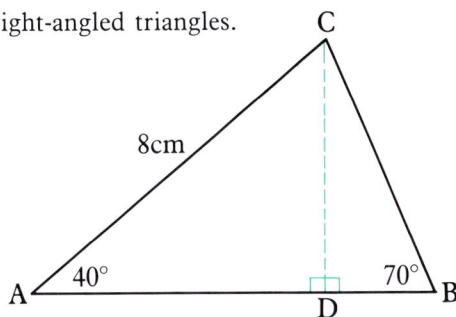

Step 2 Calculate the length of the altitude (in this case CD).

$$\sin A = \frac{CD}{AC}$$

$$\sin 40° = \frac{h}{8}$$

$$0.643 = \frac{h}{8}$$

$$h = 0.643 \times 8 = 5.144$$

CD = 5.14 cm (to 2 decimal places)

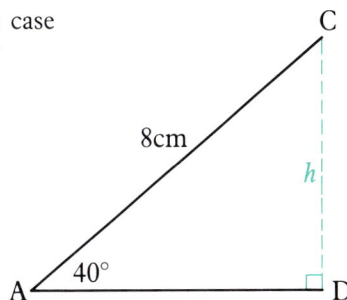

Step 3 Calculate the length of BC.

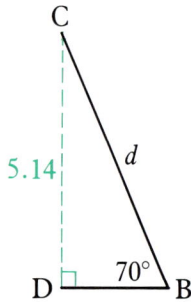

$$\sin B = \frac{CD}{BC}$$

$$\sin 70° = \frac{5·14}{d}$$

$$0·940 = \frac{5·14}{d}$$

$$0·940 \times d = 5·14$$

$$d = \frac{5·14}{0·940} = 5·47 \quad \text{(to 2 decimal places)}$$

$$BC = 5·47 \text{ cm}$$

Try these

5 Find the length of PR.

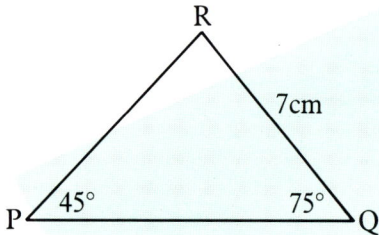

6 Find the length of BC.

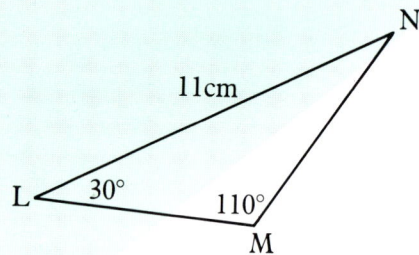

7 Find the length of NM.

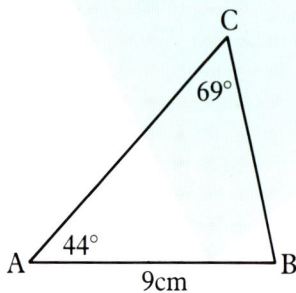

Note how you make two right-angled triangles.

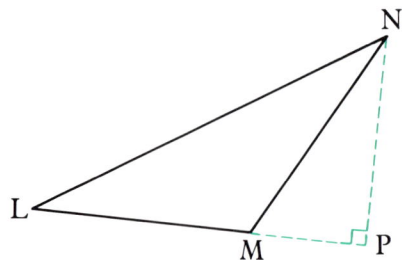

8 Find the length of AB.

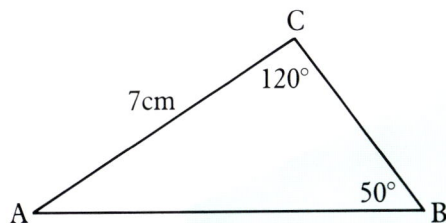

9 Find the size of the angle at Q.

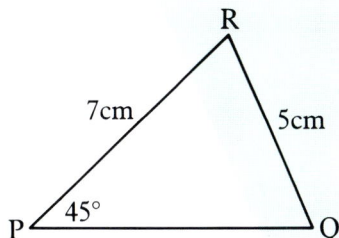

10 Find the size of the angle at C.

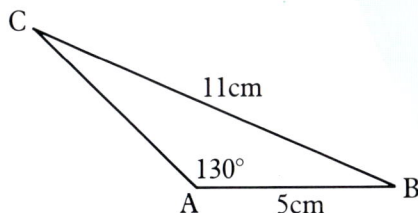

11 Find
  (i) the length of QR,
  (ii) the length of PQ (Pythagoras Theorem might help),
  (iii) the size of the angle at R.

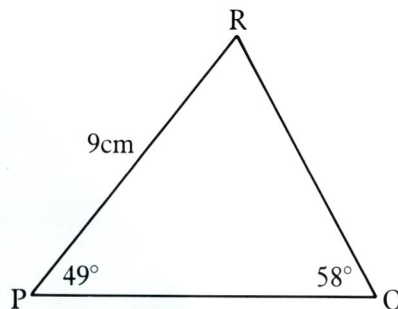

12 In answering question 5–11,
  (i) What was the minimum amount of information required;
  (ii) which trig ratio was used;
  (iii) which length was calculated in every case?

13 Can you find the length of BC?

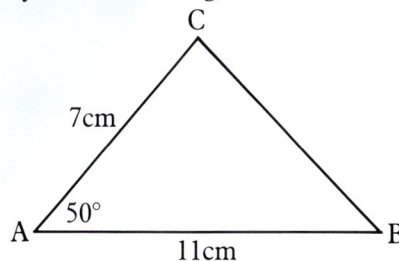

The method used in questions 5–11 does not work.

How does the information provided for the triangle in question 13 differ from that given in questions 5–11?

We shall return to examples like this one at a later stage in the unit.

The sine rule

It is possible to 'short cut' the method used in the above example. Some of the steps can be missed out. Let us see why. We will examine two situations: an acute-angled and an obtuse-angled triangle.

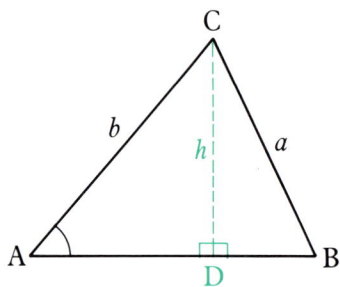

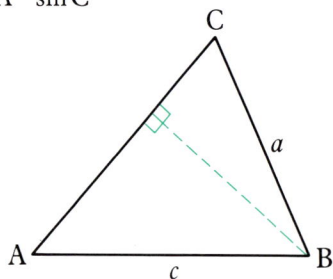

Angle at A is acute

In triangle ACD

$$\sin A = \frac{h}{b}$$

So $h = b \sin A$

Angle at A is obtuse

In triangle ACD

$$\sin(180° - A) = \frac{h}{b}$$

But $\sin(180° - A) = \sin A$

So $\sin A = \frac{h}{b}$

So $h = b \sin A$

In triangle BCD

$$\sin B = \frac{h}{a}$$

So $h = a \sin B$

In triangle BCD

$$\sin B = \frac{h}{a}$$

So $h = a \sin B$

In both cases

$$h = b \sin A$$
$$h = a \sin B$$

So $b \sin A = a \sin B$

(This avoids having to calculate the altitude).

$$b \sin A = a \sin B$$

becomes $\dfrac{b \sin A}{\sin A \ \sin B} = \dfrac{a \sin B}{\sin A \ \sin B}$ (by dividing each side by $\sin A \ \sin B$)

So $\dfrac{b}{\sin B} = \dfrac{a}{\sin A}$

or $\dfrac{a}{\sin A} = \dfrac{b}{\sin B}$

By starting with a different altitude, you could prove that

$$\frac{a}{\sin A} = \frac{c}{\sin C}$$

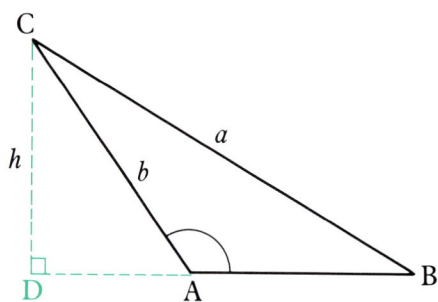

In any triangle ABC, $\dfrac{a}{\sin A} = \dfrac{b}{\sin B} = \dfrac{c}{\sin C}$

9

This is called the **sine rule**.

EXAMPLE

Calculate the length of RQ.

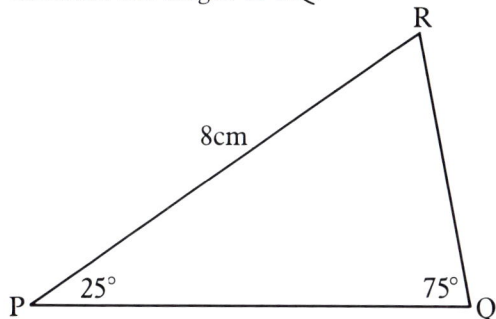

Step 1 Decide which information you have for the sine rule.

$$\dfrac{p}{\sin P} \overset{(?)}{\checkmark} = \dfrac{q}{\sin Q} \checkmark = \dfrac{r}{\sin R}$$

So use $\dfrac{p}{\sin P} = \dfrac{q}{\sin Q}$

Step 2 $\dfrac{p}{\sin 25} = \dfrac{8}{\sin 75}$

$\dfrac{p}{0 \cdot 423} = \dfrac{8}{0 \cdot 966}$

$p = \dfrac{8 \times 0 \cdot 423}{0 \cdot 966}$

$p = 3 \cdot 50$ (to 2 decimal places)

RQ = 3·50 cm

Try these

In solving the following questions, use the sine rule if you can.
Otherwise, use the method of forming two right-angled triangles.

14 Calculate the length of AB.

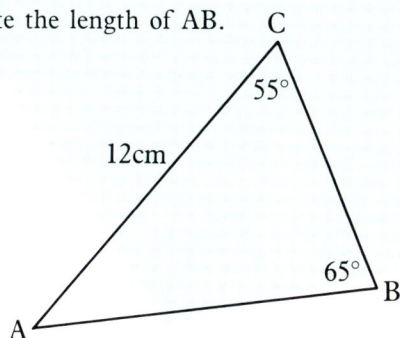

15 Find the length of QR.

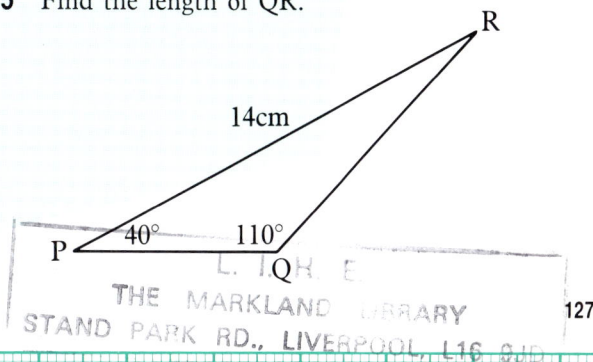

16 Find the length of AB. (Remember that the sum of the angles of a triangle is 180°.)

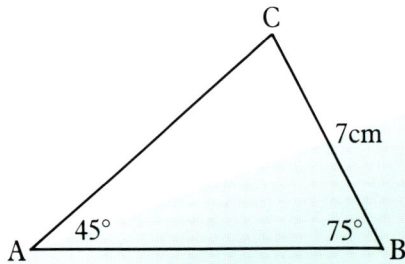

17 Triangle LMN is isosceles. Find the length of the two equal sides.

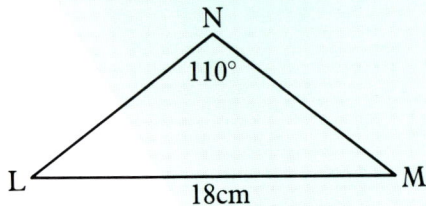

18 Calculate the size of the angle at P.

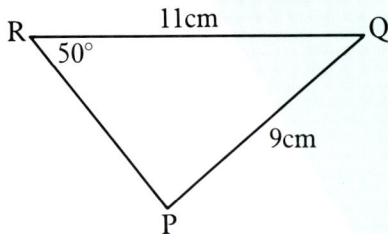

19 Calculate the size of the angle at C.

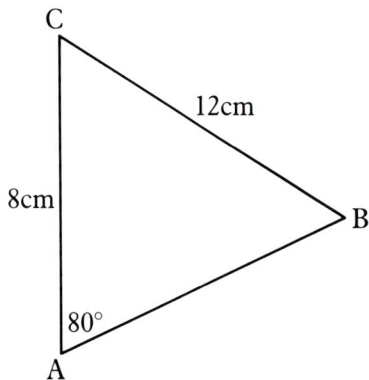

20 A racing course for a boardsailing competition is approximately an isosceles triangular course marked by three buoys.

The triangle is sailed first, followed by the 'sausage' section.

The longest section of the course is approximately 700 m and the shorter sections 500 m.

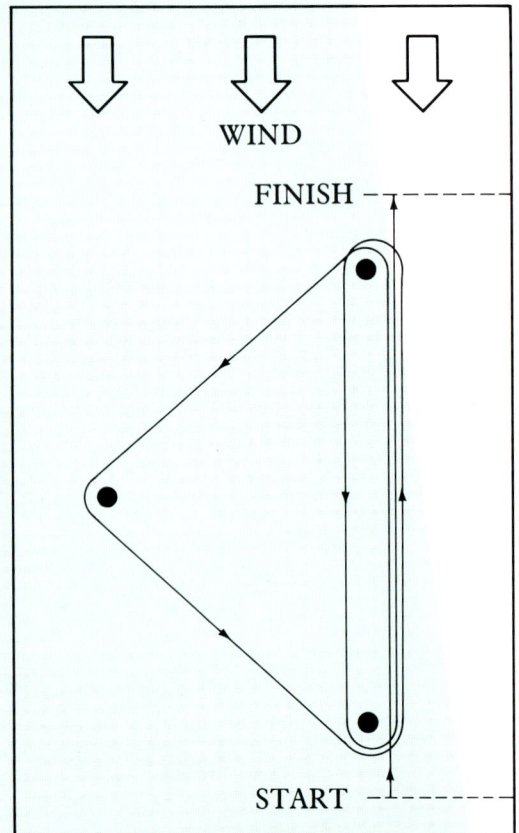

(a) Prepare a plan to show where the buoys should be placed. Include the angles.
(b) The start and finish lines are 20 m away from the nearest buoy. Calculate the length of a boardsailing course.

9

Let us have another attempt at trying to solve question 13.

EXAMPLE

Calculate the length of BC (question 13).

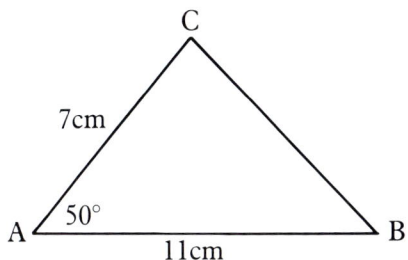

Step 1   Form two right-angled triangles.

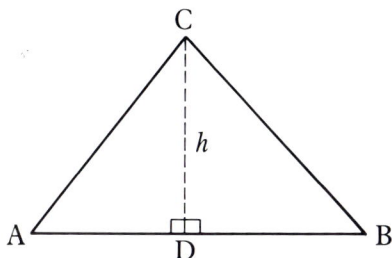

Step 2   Calculate the length of CD.

$$\cos 50° = \frac{AD}{AC}$$

$$AD = 7 \times \cos 50°$$
$$AD = 4·50 \text{ cm}\quad\text{(to 2 decimal places)}$$

Now use Pythagoras:
$$AC^2 = AD^2 + CD^2$$
$$7^2 = 4·5^2 + h^2$$
$$h = 5·36 \text{ cm}\quad\text{(to 2 decimal places)}$$

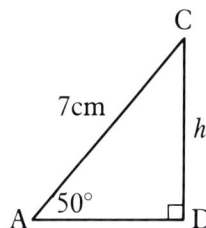

Step 3   Calculate the length of BC.

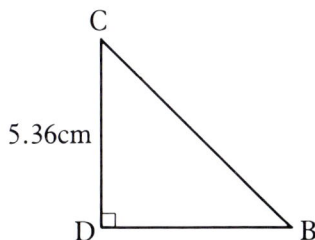

$$BD = AB - AD$$
$$BD = 11 - 4·5 = 6·5 \text{ cm}$$

Again, using Pythagoras:
$$BC^2 = CD^2 + BD^2$$
$$= 5·36^2 + 6·5^2$$

$$BC = 8·42 \text{ cm}\quad\text{(to 2 decimal places)}$$

21 Calculate the length of QR.

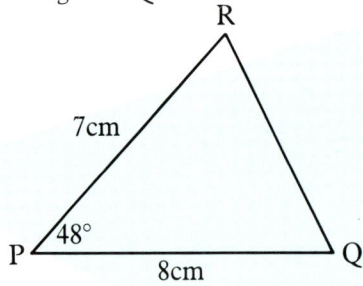

22 Calculate the length of LN.

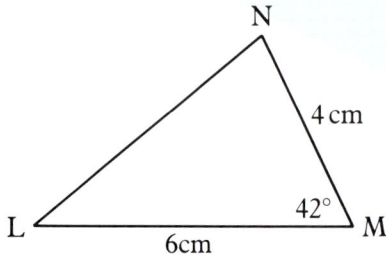

23 Calculate the length of BC.

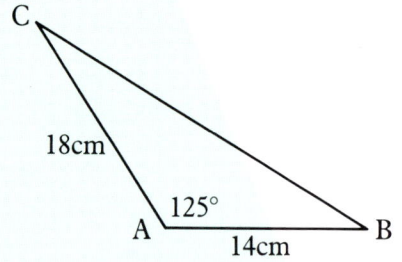

Note how you make two right-angled triangles.

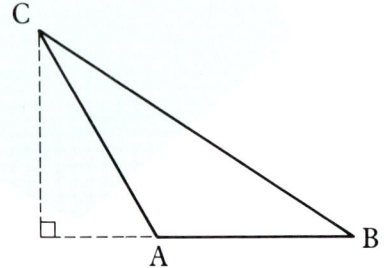

The cosine rule

It is possible to 'short cut' the method used in the last example.

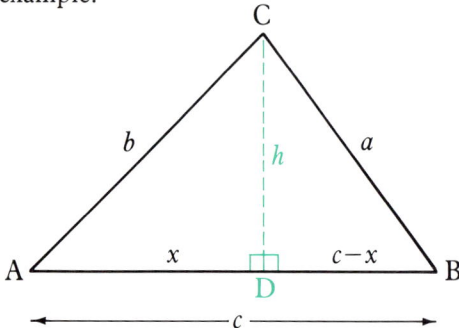

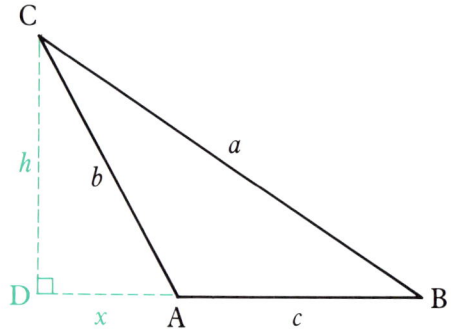

Angle at A is acute

Using Pythagoras Theorem in triangle BCD.

$a^2 = h^2 + (c - x)^2$
$a^2 = h^2 + c^2 + x^2 - 2cx$
$a^2 = (h^2 + x^2) + c^2 - 2cx$
$\quad\ b^2 = h^2 + x^2$
$a^2 = b^2 + c^2 - 2cx$

$\dfrac{x}{b} = \cos A; \quad x = b \cos A$

So
$a^2 = b^2 + c^2 - 2bc\cos A$

Angle at A is obtuse

Using Pythagoras Theorem in triangle BCD.

$a^2 = h^2 + (c + x)^2$
$a^2 = h^2 + c^2 + x^2 + 2cx$
$a^2 = (h^2 + x^2) + c^2 + 2cx$
$\quad\ b^2 = h^2 + x^2$
$a^2 = b^2 + c^2 + 2cx$

$\dfrac{x}{b} = \cos(180 - A) = -\cos A; \quad x = -b \cos A$

So
$a^2 = b^2 + c^2 - 2bc\cos A$

In any triangle ABC, $a^2 = b^2 + c^2 - 2bc\cos A$

This is called the **cosine rule**.

9

Try these

24 Using the cosine rule, complete
$p^2 = q^2 + \ldots \ldots$

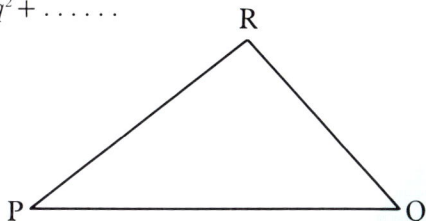

25 Using the cosine rule, complete
$b^2 = \ldots \ldots$

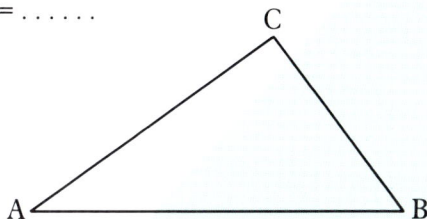

26 Calculate the length of QR.

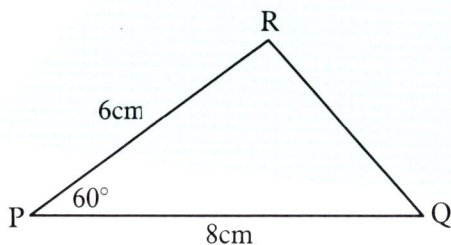

27 Calculate the length of LN.

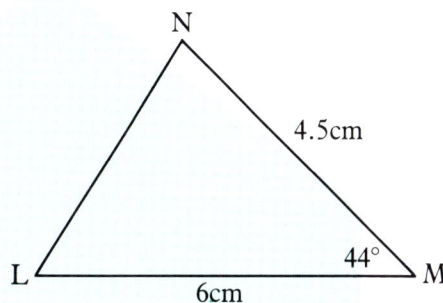

28 Calculate the length of AB.

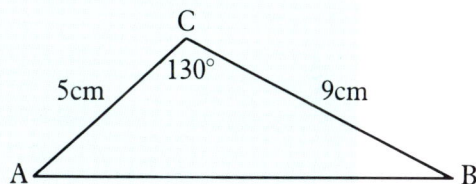

The cosine rule

When calculating an angle of a triangle given the lengths of the 3 sides.

To find the size of an angle A
$a^2 = b^2 + c^2 - 2bc\cos A$
$2bc\cos A = b^2 + c^2 - a^2$
$$\cos A = \frac{b^2 + c^2 - a^2}{2bc}$$

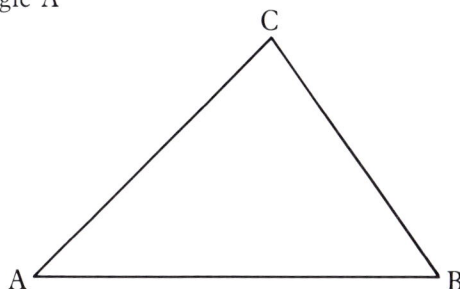

This is the cosine rule in a different form.

29 Use the formula established above to find the size of Angle A.

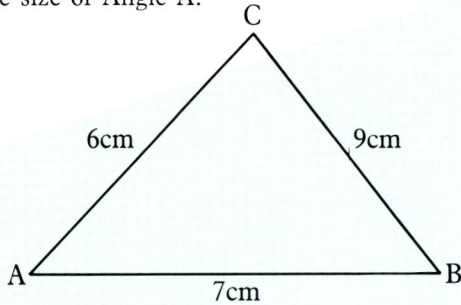

30 Using the cosine rule, complete

$$\cos B = \frac{a^2 + \ldots\ldots}{\ldots\ldots}$$

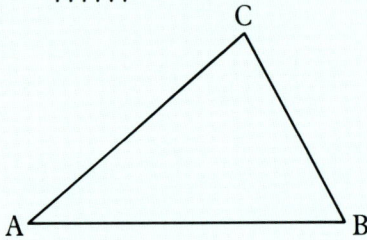

31 Using the cosine rule, complete
$\cos R =$

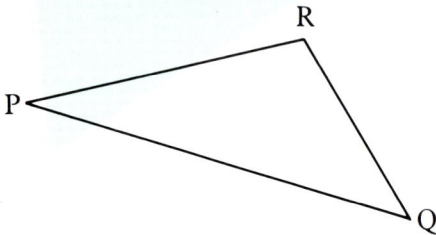

32 Calculate the size of angle N.

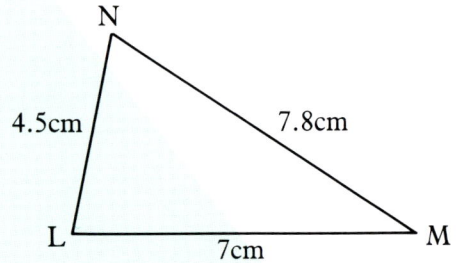

33 Calculate the size of angle R. Remember that cosines of angles between 90° and 180° are negative.

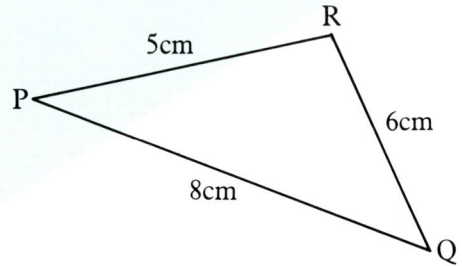

34 Calculate the largest angle in triangle ABC.

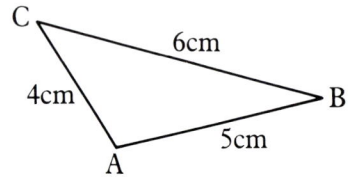

Sine Rule or Cosine Rule

Select the correct rule and complete the following questions:

35 Calculate the length of PR.

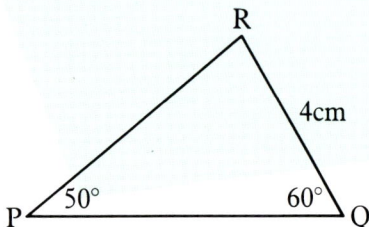

36 Calculate the size of angle C.

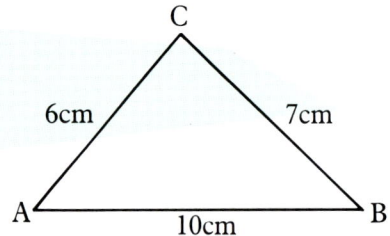

37 Calculate the size of angle L.

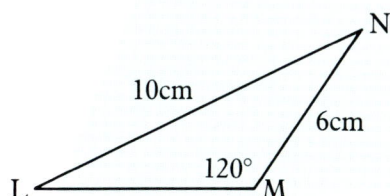

38 Calculate the length of QR.

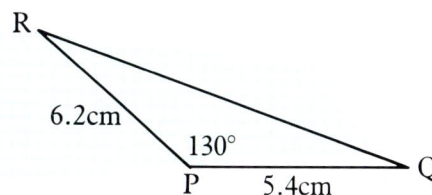

The area of a triangle

You will recollect how to calculate the area of a triangle.
Area of triangle $=\frac{1}{2}\times$ length of base $\times$ height

$$\text{area}=\tfrac{1}{2}\times AB\times DC$$

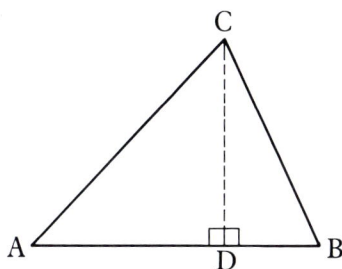

In the *Yellow Book*, you were given the following question
as an investigation. Try it again.

Try this

39 In triangle PQR, $p=10$ cm, $r=8$ cm, angle Q$=40°$. Calculate the
area. (*Hint:* You have to use the trig ratios to calculate the
lengths of the sides you require.)

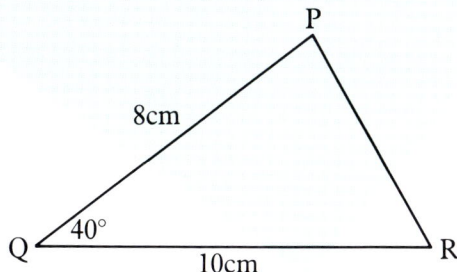

Let us see if it is possible to find a formula which is a
useful short cut to the answer.

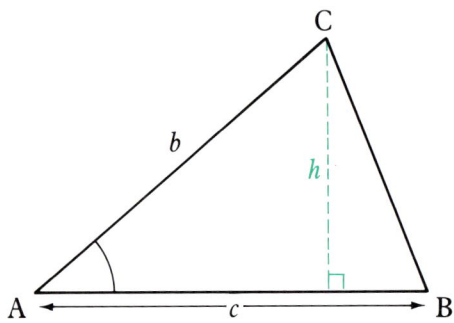

Angle A is acute

Area of triangle ABC $= \frac{1}{2}ch$
$h = b\sin A$

Area of triangle ABC $= \frac{1}{2}bc\sin A$

Angle A is obtuse

Area of triangle ABC $= \frac{1}{2}ch$
$h = b\sin(180 - A) = b\sin A$

Area of triangle ABC $= \frac{1}{2}bc\sin A$

The area of triangle ABC $= \frac{1}{2}bc\sin A$

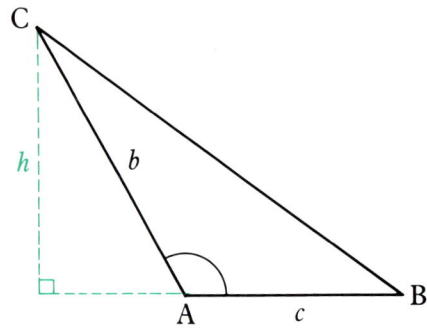

You could also prove that
area of triangle ABC $= \frac{1}{2}ab\sin C = \frac{1}{2}ac\sin B$

Try these

40 Calculate the area of each triangle.

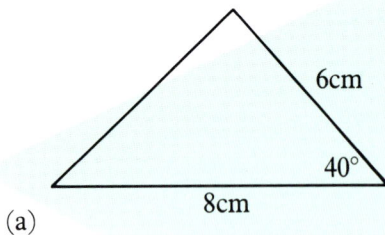

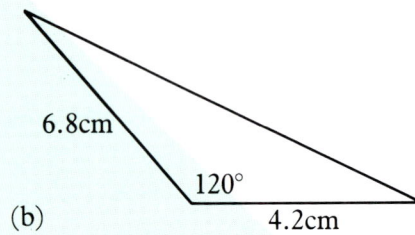

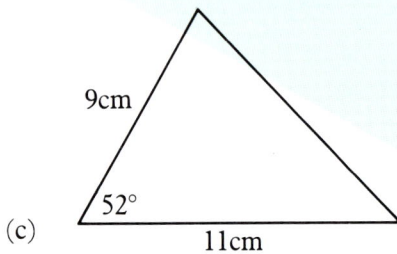

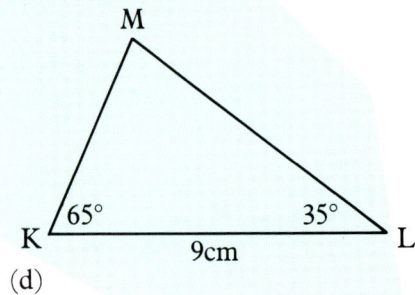

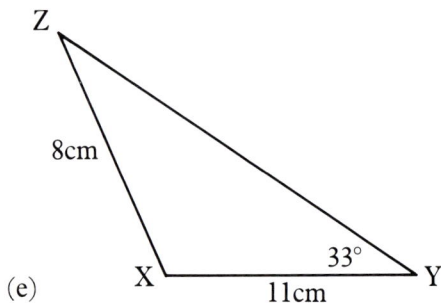

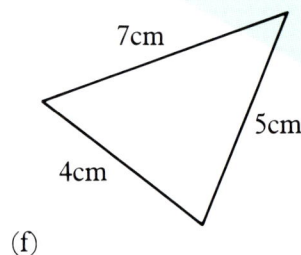

(a)

(b)

(c)

(d)

(e)

(f)

41 Can you find formulae for calculating the area of a parallelogram and a kite?

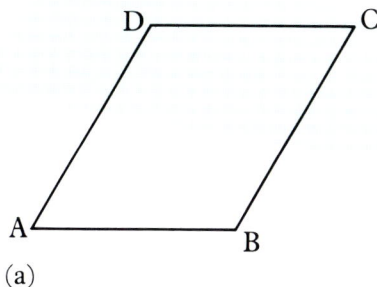

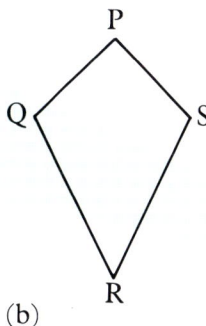

9

(a)

(b)

Trig ratios — in 3-dimensions

The difficulty in using trig ratios in 3-dimensional objects (e.g. cuboid) is selecting the triangle (or triangles) you require. Your teacher may have some equipment that might help you to understand this section.

EXAMPLE

A, B, C, D, E, F, G, H is a cuboid of length 10 cm, height 6 cm and width 5 cms.
Calculate the size of angle BGC.

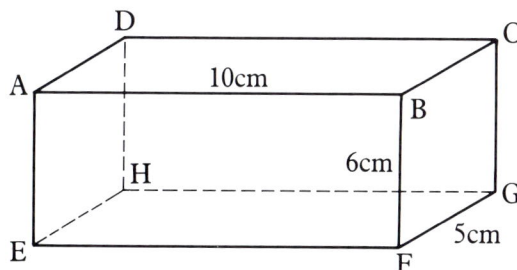

Step 1 Select the plane that contains the angle.

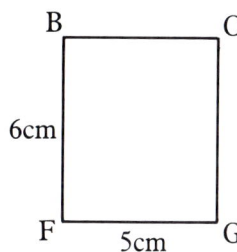

Step 2 Draw the appropriate triangle showing all known information.

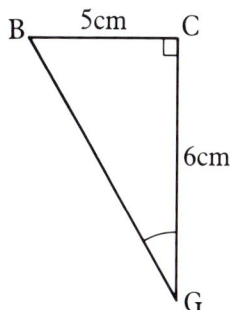

Step 3 Complete calculation

$$\tan G = \frac{BC}{CG}$$

$$\tan G = \frac{5}{6}$$

$$G = 39 \cdot 8°$$

angle BGC $= 39 \cdot 8°$

42 K, L, M, N, P, Q, R, S is a cube of side 6 cm.

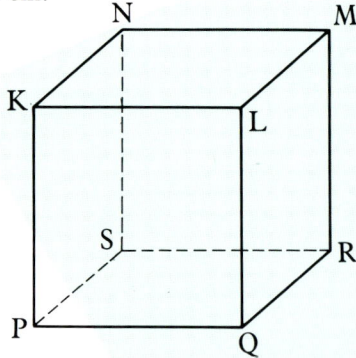

(a) Calculate the length of PL.
(b) Calculate the size of angle MSN.

43 A, B, C, D, E, F, G, H is a cuboid of length 12 cm, height 6 cm and width 8 cm.

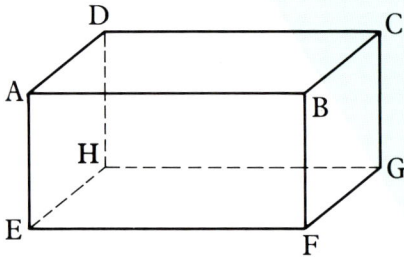

(a) Calculate the length of AC.
(b) Calculate the size of angle HFG.
(c) Calculate the size of angle BHF.

44 A, B, C, D, K, L, M, N is a cube of side 8 cm. S is the mid-point of side KL.

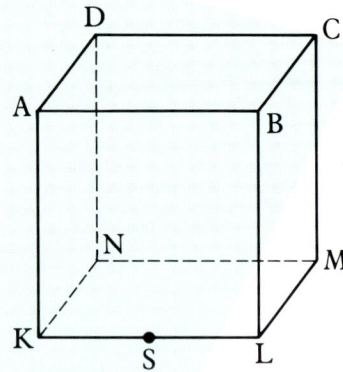

(a) Calculate the size of angle ASB.
(b) Calculate the size of angle CSM.

45 P, A, B, C, D is a pyramid with a square base of side 30 cm.
The vertical height of the pyramid is 40 cm.
Calculate the size of angle PBC.

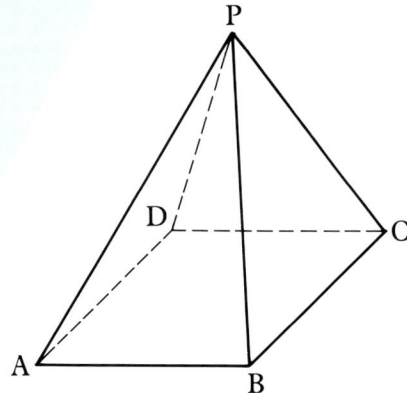

Trigonometric equations

RELATIONSHIP BETWEEN SINE, COSINE AND TANGENT

Let us return to the spinning disc we met in unit 7. This time let the radius of the disc be r.

$$\sin A = \frac{h}{r} \qquad \cos A = \frac{d}{r} \qquad \tan A = \frac{h}{d}$$

$$\frac{\sin A}{\cos A} = \frac{\frac{h}{r}}{\frac{d}{r}} = \frac{h}{d} = \tan A$$

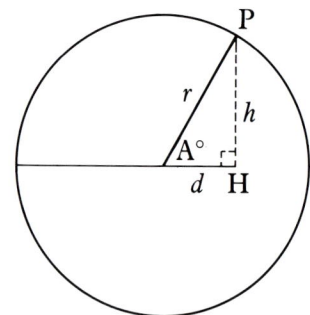

$$\tan A = \frac{\sin A}{\cos A} \quad \text{so long as } \cos A \neq 0$$

RELATIONSHIP BETWEEN SINE AND COSINE

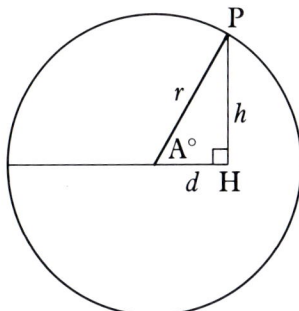

$$h^2 + d^2 = r^2 \qquad \text{(Pythagoras Theorem)}$$

$$\frac{h^2}{r^2} + \frac{d^2}{r^2} = 1 \qquad \text{(dividing both sides by } r^2)$$

$$\left(\frac{h}{r}\right)^2 + \left(\frac{d}{r}\right)^2 = 1$$

$$(\sin A)^2 + (\cos A)^2 = 1$$

This is usually written as

$$\mathbf{\sin^2 A + \cos^2 A = 1}$$

Note: $\sin^2 A$ represents $(\sin A)^2 = \sin A \times \sin A$

The above two formulae have only been proved for the first quadrant but they can easily be extended to the remaining quadrants.

Check these formulae using your calculator. Select values of A for $0 \leq A \leq 360$.

Try these

46 Solve the following equations to the nearest degree for $0 \leq x \leq 360$.

(a) $\tan x^\circ = 1$
(b) $\sin x^\circ = \sin 140^\circ$
(c) $\sin x^\circ + \sin 50^\circ = 0$
(d) $\cos x^\circ - \cos 150^\circ = 1$
(e) $3 \sin x^\circ = 1 \cdot 5$
(f) $\sin^2 x^\circ = 0 \cdot 36$
(g) $\cos x^\circ = -\sin 120^\circ$

47 Express each of the following in terms of a single ratio:

(a) $\dfrac{\sin 50^\circ}{\cos 50^\circ}$ (b) $\cos A \times \tan A$

48 Find the value of $\sin^2 220^\circ + \cos^2 220^\circ$

49 (a) Prove that $4\cos^2 A = 4 - 4\sin^2 A$
(b) Express $1 - 3\sin^2 A$ in terms of $\cos^2 A$

50 Prove that $\tan A \sin A = \dfrac{1 - \cos^2 A}{\cos A}$

Now try this...

A A mountaineering party normally has a 40 m rope with them. A major hazard for such a party can be crossing a stream in spate when there is no nearby bridge. The rope can be used to protect the first member of the party to wade across the stream.

One method is called 'the continuous loop system'. The ends of the rope are tied together. A 40 m rope should enable you to cross a stream 12 m wide. The system is as follows:

C should be one of the larger and stronger members of the party. C sets off across and slightly downstream supporting himself on the rope held by A. In the event of C slipping in, C is pulled to the shore by B.

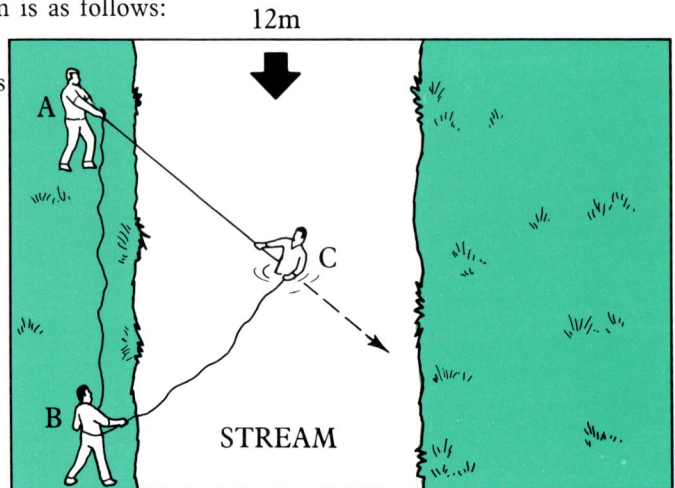

What is the maximum distance that A and B should stand apart if the stream is 12 m wide and all the rope is used in crossing it?

B Let us return to the orienteerer at the start of this unit.

The direct route from control point 1 to control point 2 is through fairly dense forest. The average speed for the orienteerer over this terrain is 9 km/hr.

The distance between the control points is 1·2 km.

The bearing of control point 2 from control point 1 is 290°.

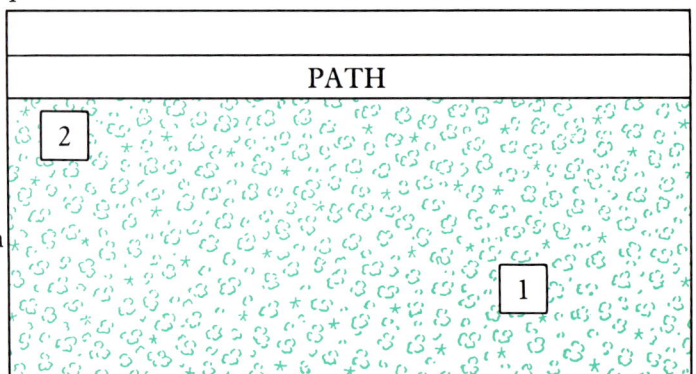

A path (just beyond control point 2) runs approximately East–West. When running along a path, an orienteerer can average 14 km/hr.

Should the orienteerer head towards the path and run part of the way along the path?

Can you plan the route that would be the fastest to take?

KEY QUESTIONS

K1 (a) Calculate the length of BC.

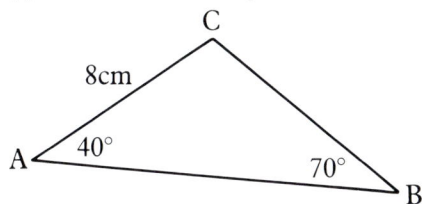

(b) Calculate the size of angle Q.

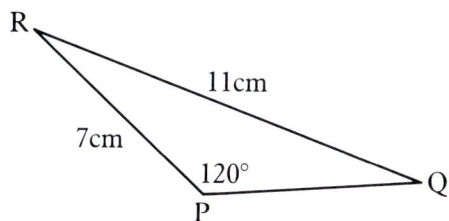

K2 (a) Calculate the size of the largest angle.

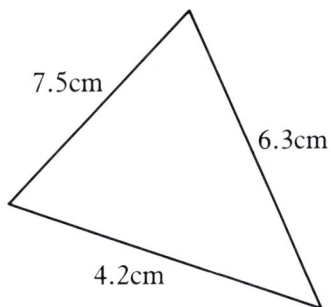

(b) Calculate the length of PR.

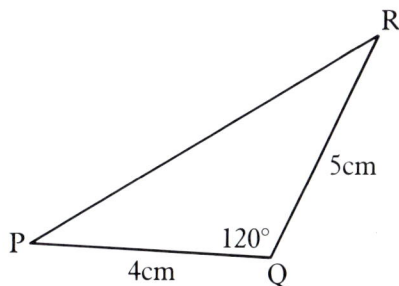

K3 (a) Calcluate the area of triangle ABC.

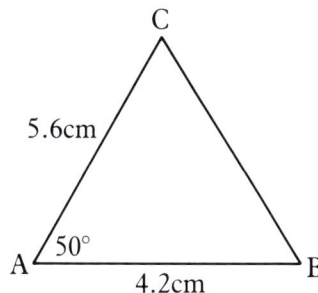

(b) Calculate the area of triangle PQR.

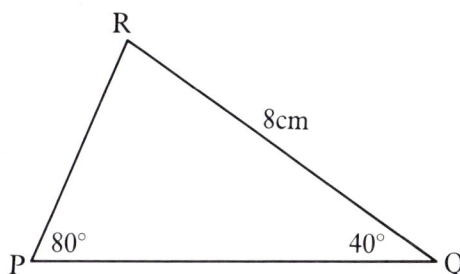

K4 (a) Solve the following equations to the nearest degree for $0 \leqslant x \leqslant 360$.
   (i) $\sin x° = 0.5$
   (ii) $\sin x° = -\cos 120°$

(b) Simplify $\dfrac{\tan A}{\sin A}$.

TRANSFORMATIONS AND MATRICES

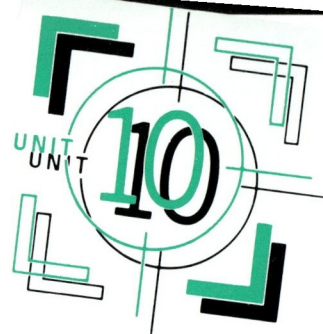

This pattern of triangles has been produced by using transformations such as half-turn, quarter-turn, reflection in the line $y = x$, reflection in the x-axis

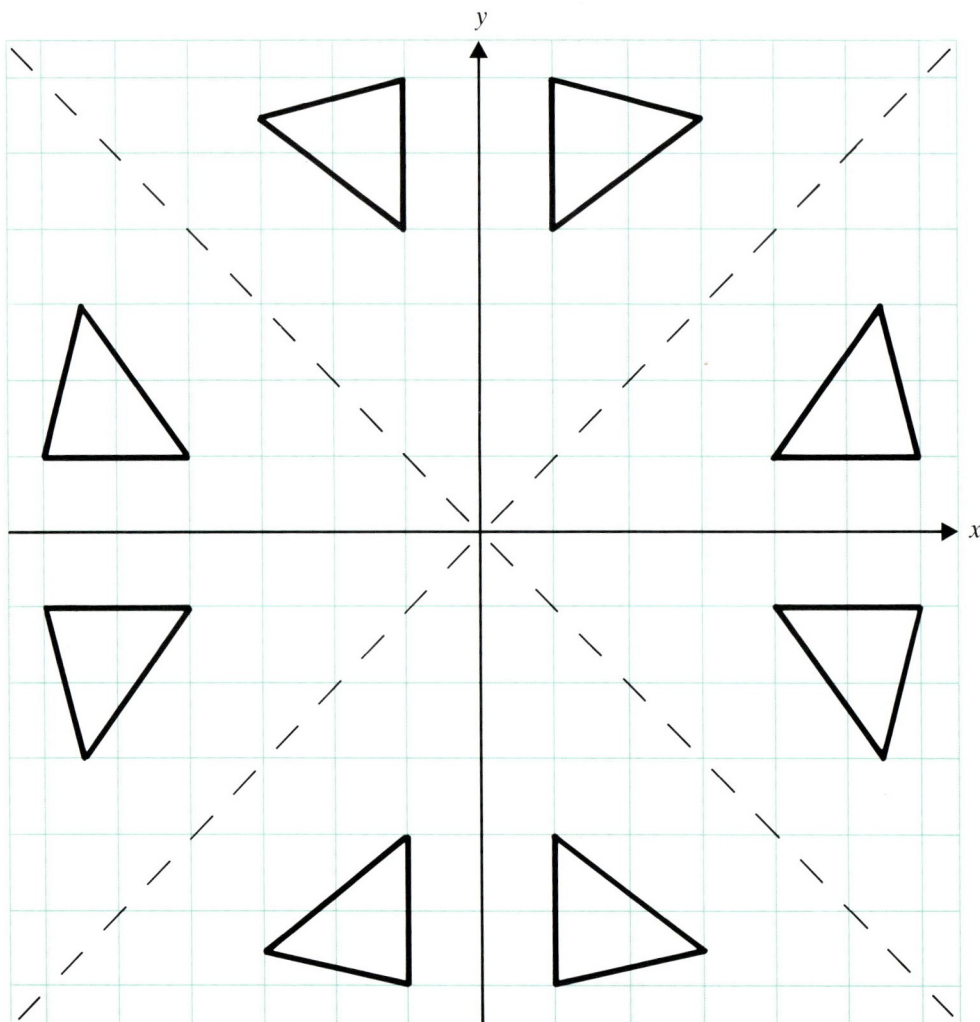

In this unit, we will use equations and matrices to describe the effect of these transformations, singly and in combination.

We will, for example, be able to state which single transformation has the same effect as reflection in the line $y = x$ followed by reflection in the x-axis.

Perhaps you can guess the answer by looking at the diagram.

In a unit called Symmetry in the *Red Book*, we asked you to investigate the reflection of a point $P(x, y)$ in the line $y = x$. If you tried that investigation, what conclusion did you reach?

Let's look at a few examples.

$(2,3) \longrightarrow (3,2)$

$(-4,1) \longrightarrow (1,-4)$

$(-2,-5) \longrightarrow (-5,-2)$

$(3,-1) \longrightarrow (-1,3)$

$(0,4) \longrightarrow (4,0)$

$(2,0) \longrightarrow (0,2)$

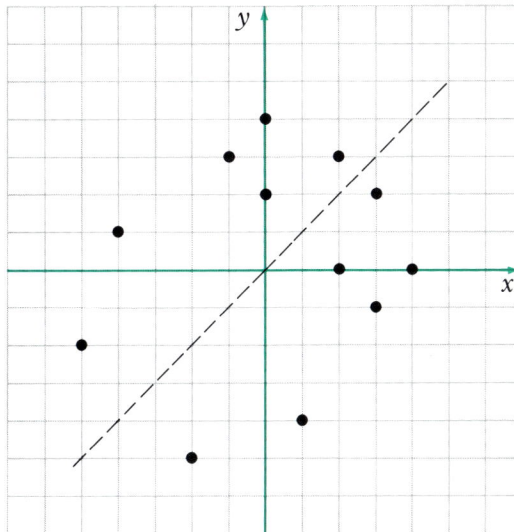

From the above, it seems reasonable to make the *general* statement:

If $P(x, y)$ is reflected in the line $y = x$ its image is $P'(y,x)$

We could make the statement in a different way

If P' is the reflection of a point P in the line $y = x$, the x-coordinate of P' is the same as the y-coordinate of P and the y-coordinate of P' is the same as the x-coordinate of P.

We could also write the statement in terms of equations in this way:

If (x', y') is the reflection of (x, y) in the line $y = x$,
$x' = y$
$y' = x$

Try this

1 Work out the versions of the above three statements for reflection in the line $y = -x$.

Let's now consider the reflection of a point in *any* line that passes through the origin (0,0).

In the diagram, P(x, y) is reflected in the line OL, which makes an angle a with Ox, giving an image P'(x', y')

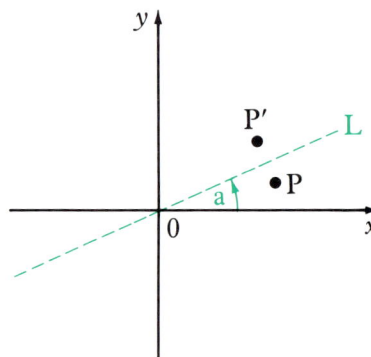

It can be shown that
$$x' = (\cos 2a)x + (\sin 2a)y$$
$$y' = (\sin 2a)x - (\cos 2a)y$$

Let's check this out for two special cases.

Reflection in the line $y = x$

$a = 45°$

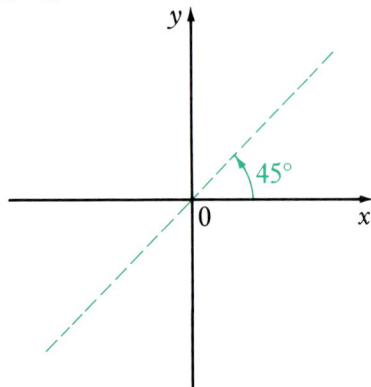

Reflection in the x-axis

$a = 0°$

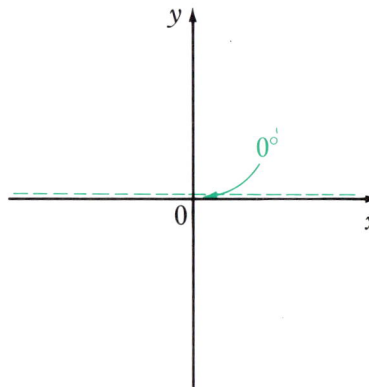

$2a = 90°$
$\sin 2a = 1$
$\cos 2a = 0$

$x' = (\cos 2a)x + (\sin 2a)y$
$\quad = 0x + 1y$
$\quad = y$

$y' = (\sin 2a)x - (\cos 2a)y$
$\quad = 1x - 0y$
$\quad = x$

reflection in $y = x$

$x' = y$
$y' = x$

$2a = 0°$
$\sin 2a = 0$
$\cos 2a = 1$

$x' = (\cos 2a)x + (\sin 2a)y$
$\quad = 1x + 0y$
$\quad = x$

$y' = (\sin 2a)x - (\cos 2a)y$
$\quad = 0x - 1y$
$\quad = -y$

reflection in the x-axis

$x' = x$
$y' = -y$

We see that the equations that we get here are the same as those that we got previously by looking at diagrams.

Try this

2 (a) Use the equations for reflection in a line at angle a to Ox,

$$x' = (\cos 2a)x + (\sin 2a)y, \qquad y' = (\sin 2a)x - (\cos 2a)y,$$

to complete the following.

The line $y = -x$

$a = ?°$

$2a = ?° \quad \sin 2a = ? \quad \cos 2a = ?$

$\begin{aligned} x' &= (\cos 2a)x + (\sin 2a)y & \qquad y' &= (\sin 2a)x - (\cos 2a)y \\ &= ?x + ?y & &= ?x - ?y \\ &= ? & &= ? \end{aligned}$

Are your results the same as you got previously for reflection in the line $y = -x$?

(b) Obtain a pair of equations for x' and y', the coordinates of the image of (x, y) after reflection in the y-axis.

In the Symmetry unit in the *Red Book*, we also asked you to investigate the reflection of a point $P(x, y)$ in the lines $x = k$ and $y = j$.

You should be able to see from the diagram that (x', y'), the image of (x, y) in the line $x = k$, is given by the equations
$x' = 2k - x \qquad [x' = x + (k - x) + (k - x)]$
$y' = y$

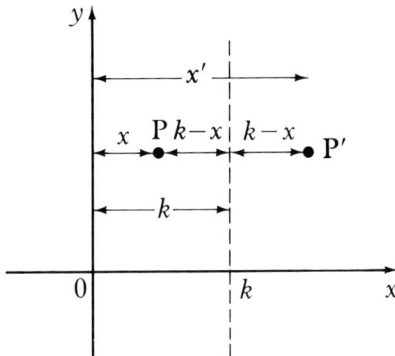

reflection in $y = x$, reflection in $x = 5$, reflection in $y = -3$ are all examples of **transformations** of points (and shapes).

Other examples of transformations that we have met previously are rotations and translations. We will look at these later in this unit.

Try these

3 Write down a pair of equations for the transformation: *reflection in the line $y = j$.*

4 Use the equations that we have obtained so far to write down (no diagrams!) the images of the given points under the following transformations:

(a) *reflection in y = x* $(-10, 50) \longrightarrow ?$
(b) *reflection in y = 0* $(40, -20) \longrightarrow ?$
(c) *reflection in y = 3* $(-20, -50) \longrightarrow ?$
(d) *reflection in y = -x* $(40, -10) \longrightarrow ?$
(e) *reflection in x = 0* $(-30, 50) \longrightarrow ?$
(f) *reflection in x = -4* $(10, 10) \longrightarrow ?$

Let's look now at another transformation, **rotation**.

You may remember, from the Symmetry unit, examples like the following.

Show P′, the image of P, after a 180° rotation about O (half-turn), e.g. $P(2, -3) \longrightarrow P'(-2, 3)$

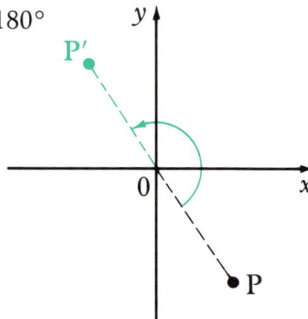

Show Q, the image of Q, after a 90° anticlockwise rotation about O (quarter-turn) e.g. $Q(-1, 4) \longrightarrow Q'(-4, -1)$

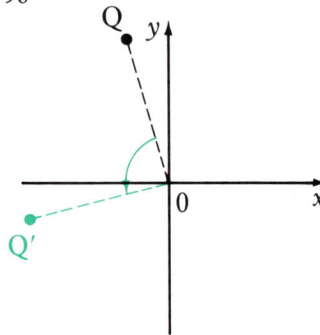

Show R′, the image of R, after a 90° clockwise rotation about O (quarter-turn) e.g. $R(-3, -1) \rightarrow R'(-1, 3)$

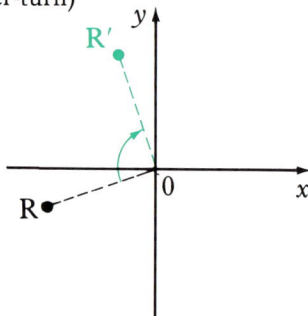

Try this

5 By using more examples like those above, try to obtain three pairs of equations for the three transformations *half-turn, quarter-turn anticlockwise, quarter-turn clockwise.*

In this diagram, P(x, y) is rotated through angle a about O giving an image P$'(x', y')$

It can be shown that
$x' = (\cos a)x - (\sin a)y$
$y' = (\sin a)x + (\cos a)y$

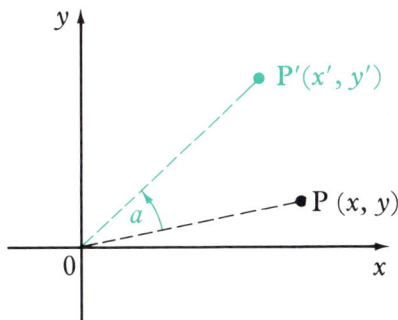

Let's check this out for a special case that we looked at earlier.

Half-turn
$a = 180°$ $\cos a = -1$
$\qquad\qquad \sin a = 0$

$x' = (\cos a)x - (\sin a)y \qquad y' = (\sin a)x + (\cos a)y$
$\quad = -1x - 0y \qquad\qquad\quad = 0x + (-1)y$
$\quad = -x \qquad\qquad\qquad\quad = -y$

$x' = -x$
$y' = -y$

The result here should be the same as that in Try This 5.

Try these

6 Use the equations for rotation through an angle a about O

$x' = (\cos a)x - (\sin a)y$
$y' = (\sin a)x + (\cos a)y$

to obtain equations for x' and y', the coordinates of the image of (x, y) after the following rotations about O

(a) 90° anticlockwise
(b) 90° clockwise $(-90°)$

Are your results the same as you got previously (in Try This 5)?

7 Use the equations that we have obtained to write down (again, no diagrams) the images of the given points under the following rotations about O.

(a) 180°: $(-1, 5) \longrightarrow$?
(b) 90°: $(4, -2) \longrightarrow$?
(c) $-90°$: $(-2, -5) \longrightarrow$?
(d) half-turn: $(4, -1) \longrightarrow$?
(e) quarter-turn anticlockwise: $(-3, 5) \longrightarrow$?
(f) quarter-turn clockwise: $(1, 1) \longrightarrow$?

Another transformation that we looked at in a previous book
was **translation**.

Each point in the diagram has been given

a translation $\begin{pmatrix} -2 \\ 5 \end{pmatrix}$.

$(3,1) \longrightarrow (1,6)$
$(-1,-4) \longrightarrow (-3,1)$
$(-2,3) \longrightarrow (-4,8)$
$(4,-3) \longrightarrow (2,2)$

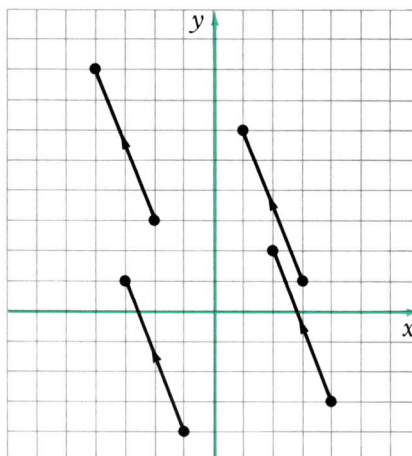

It seems obvious that

if (x', y') is the image of (x, y) after a translation $\begin{pmatrix} a \\ b \end{pmatrix}$,

$x' = x + a$
$y' = y + b$

Try this

8 Translations T_1 and T_2 are as follows

$$T_1 = \begin{pmatrix} 1 \\ -3 \end{pmatrix} \qquad T_2 = \begin{pmatrix} -2 \\ 4 \end{pmatrix}$$

Write down the images of the given points after the following
translations,

(a) $(-1,5)$ after T_1
(b) $(4,-2)$ after T_2
(c) $(-2,-5)$ after T_1
(d) $(4,-1)$ after T_2
(e) $(-3,5)$ after T_1 followed by T_2
(f) $(-3,5)$ after T_2 followed by T_1

We looked, in the *Yellow Book*, at **enlargements and
reductions**.

This diagram shows an enlargement,
centre O, scale factor 2, of the rectangle
ABCD.

$A(3,2) \longrightarrow A'(6,4)$
$B(-3,2) \longrightarrow B'(-6,4)$
$C(-3,-2) \longrightarrow C'(-6,-4)$
$D(3,-2) \longrightarrow D'(6,-4)$

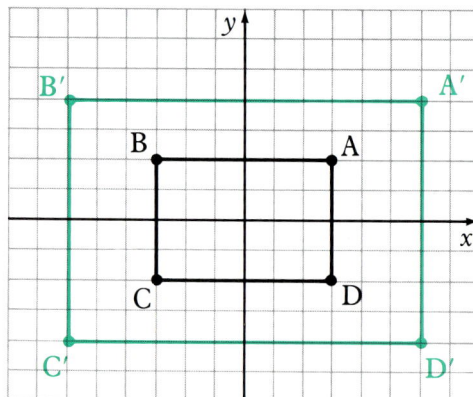

This diagram shows a reduction, centre O, scale factor $\frac{1}{2}$ of the triangle PQR.

$P(6,4) \longrightarrow P'(3,2)$
$Q(-4,2) \longrightarrow Q'(-2,1)$
$R(4,-4) \longrightarrow R'(2,-2)$

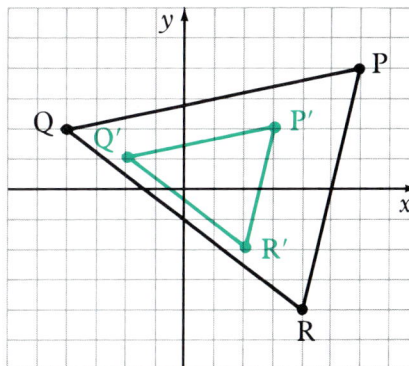

The word that is used to describe the transformation of each point of a shape after an enlargement or reduction is **dilatation**.

The dilatation is usually written as [C,k] where C is the centre of the dilatation and k is the scale factor.

If seems reasonable to say that if (x',y') is the image of (x,y) after a dilatation [O,k], then

$x' = kx$
$y' = ky$

Try these

9 Write down the images of the given points after the following dilatations.

(a) $(-1,5)$ after [O,3]

(b) $(4,-2)$ after [O,$\frac{1}{2}$]

(c) $(-2,-5)$ after [O,-2]

(d) $(4,-1)$ after [O,-0.1]

(e) $(-3,5)$ after [O,10]

(f) $(1,1)$ after [O,-0.5]

10 Consider the following transformation equations

$x' = 2x$
$y' = y$

Plot the points $(1,1)$, $(1,3)$, $(4,1)$, $(4,3)$.

Write down the images of these points after the transformation whose equations are given.

Plot the images.

Describe the transformation and give it a suitable name.

Let's summarise the transformation equations we have obtained so far.

TRANSFORMATION	EQUATIONS
reflection in the line $y=x$	$x'=y$ $y'=x$
reflection in the line $y=-x$	$x'=-y$ $y'=-x$
reflection in the x-axis	$x'=x$ $y'=-y$

TRANSFORMATION	EQUATIONS
reflection in the y-axis	$x'=-x$ $y'=y$
reflection in the line $x=k$	$x'=2k-x$ $y'=y$
reflection in the line $y=j$	$x'=x$ $y'=2j-y$
rotation about O through 180° (half-turn)	$x'=-x$ $y'=-y$
rotation about O through 90° (quarter-turn anticlockwise)	$x'=-y$ $y'=x$
rotation about O through $-90°$ (quarter-turn clockwise)	$x'=y$ $y'=-x$
translation $\begin{pmatrix} a \\ b \end{pmatrix}$	$x'=x+a$ $y'=y+b$
dilatation [O,k]	$x'=kx$ $y'=ky$

Transformations and matrices

You may recall from an earlier book a special arrangement
of numbers called an array or **matrix**.

$\begin{pmatrix} 3 & -4 \\ -2 & 5 \end{pmatrix}$ is a 2×2 matrix

$\begin{pmatrix} -2 & 1 \\ 1 & 4 \\ 7 & 0 \end{pmatrix}$ is a 3×2 matrix

$\begin{pmatrix} -1 \\ 2 \end{pmatrix}$ is a 2×1 matrix

To multiply a 2×2 matrix by a 2×1 matrix, we proceed as
follows

$$\mathbf{A} = \begin{pmatrix} a_{11} & a_{12} \\ a_{21} & a_{22} \end{pmatrix} \qquad \mathbf{B} = \begin{pmatrix} b_1 \\ b_2 \end{pmatrix}$$

(a_{11} is the entry in the first row and the first column,
a_{21} is the entry in the second row and first column,
and so on.)

$$\mathbf{AB} = \begin{pmatrix} a_{11}b_1 + a_{12}b_2 \\ a_{21}b_1 + a_{22}b_2 \end{pmatrix}$$

e.g. $\mathbf{A} = \begin{pmatrix} 3 & -4 \\ -2 & 5 \end{pmatrix} \qquad \mathbf{B} = \begin{pmatrix} -1 \\ 2 \end{pmatrix}$

$$\mathbf{AB} = \begin{pmatrix} 3 \times (-1) + (-4) \times 2 \\ (-2) \times (-1) + 5 \times 2 \end{pmatrix} = \begin{pmatrix} -11 \\ 12 \end{pmatrix}$$

Let's look at some of the transformation equations we
obtained earlier in this unit and see how matrices can help
us to use them.

For the transformation
reflection in the line $y = x$
the equations are
$x' = y$
$y' = x$

We can rewrite these equations as
$x' = 0x + 1y$
$y' = 1x + 0y$

Now consider the matrices

$$\begin{pmatrix} 0 & 1 \\ 1 & 0 \end{pmatrix} \quad \text{and} \quad \begin{pmatrix} x \\ y \end{pmatrix}$$

When we multiply these two matrices we get

$$\begin{pmatrix} 0 & 1 \\ 1 & 0 \end{pmatrix} \begin{pmatrix} x \\ y \end{pmatrix} = \begin{pmatrix} 0x + 1y \\ 1x + 0y \end{pmatrix}$$
$$= \begin{pmatrix} y \\ x \end{pmatrix}$$

and we can now write the equations using matrices as
follows

$$\begin{pmatrix} x' \\ y' \end{pmatrix} = \begin{pmatrix} 0 & 1 \\ 1 & 0 \end{pmatrix} \begin{pmatrix} x \\ y \end{pmatrix}$$

Let's do the same with another transformation;
reflection in the x-axis
$x' = x$
$y' = -y$

$x' = 1x + 0y$
$y' = 0x + (-1)y$

$$\begin{pmatrix} x' \\ y' \end{pmatrix} = \begin{pmatrix} 1 & 0 \\ 0 & -1 \end{pmatrix} \begin{pmatrix} x \\ y \end{pmatrix}$$

Try this

11 Obtain the matrix equations for the following transformations

(a) reflection in the line $y = -x$,
(b) reflection in the x-axis.

Perhaps you remember, from the work on matrices that we
went on to consider, the **unit matrix** and the **inverse
matrix** of a 2×2 matrix.

The unit matrix, $\mathbf{I} = \begin{pmatrix} 1 & 0 \\ 0 & 1 \end{pmatrix}$, is such that, if $\mathbf{X}$ is any other
matrix,
$\mathbf{IX} = \mathbf{X}$
$\mathbf{XI} = \mathbf{X}$

(In the same way, 1 is the unit number, and if x is any
other number, $1 \times x = x \quad x \times 1 = x$)

Reminder

We multiply a 2×2 matrix by a 2×2 matrix as follows:

$$\mathbf{P} = \begin{pmatrix} p_{11} & p_{12} \\ p_{21} & p_{22} \end{pmatrix} \qquad \mathbf{Q} = \begin{pmatrix} q_{11} & q_{12} \\ q_{21} & q_{22} \end{pmatrix}$$

$$\mathbf{PQ} = \begin{pmatrix} p_{11}q_{11} + p_{12}q_{21} & p_{11}q_{12} + p_{12}q_{22} \\ p_{21}q_{11} + p_{22}q_{21} & p_{21}q_{12} + p_{22}q_{22} \end{pmatrix}$$

E.g. if $\mathbf{X} = \begin{pmatrix} 1 & -3 \\ -2 & 4 \end{pmatrix}$

$$\mathbf{IX} = \begin{pmatrix} 1 & 0 \\ 0 & 1 \end{pmatrix} \begin{pmatrix} 1 & -3 \\ -2 & 4 \end{pmatrix} = \begin{pmatrix} 1 \times 1 + 0 \times (-2) & 1 \times (-3) + 0 \times 4 \\ 0 \times 1 + 1 \times (-2) & 0 \times (-3) + 1 \times 4 \end{pmatrix} = \begin{pmatrix} 1 & -3 \\ -2 & 4 \end{pmatrix} = \mathbf{X}$$

$$\mathbf{XI} = \begin{pmatrix} 1 & -3 \\ -2 & 4 \end{pmatrix} \begin{pmatrix} 1 & 0 \\ 0 & 1 \end{pmatrix} = \begin{pmatrix} 1 \times 1 + (-3) \times 0 & 1 \times 0 + (-3) \times 1 \\ (-2) \times 1 + 4 \times 0 & (-2) \times 0 + 4 \times 1 \end{pmatrix} = \begin{pmatrix} 1 & -3 \\ -2 & 4 \end{pmatrix} = \mathbf{X}$$

If $\mathbf{A}^{-1}$ is the inverse of matrix $\mathbf{A}$

$\mathbf{A}^{-1}\mathbf{A} = \mathbf{I}$

$\mathbf{A}\mathbf{A}^{-1} = \mathbf{I}$

We obtain $\mathbf{A}^{-1}$ as follows.

If $\mathbf{A} = \begin{pmatrix} a_{11} & a_{12} \\ a_{21} & a_{22} \end{pmatrix}$

$$\mathbf{A}^{-1} = \frac{1}{k} \begin{pmatrix} a_{22} & -a_{12} \\ -a_{21} & a_{11} \end{pmatrix} = \begin{pmatrix} \frac{a_{22}}{k} & -\frac{a_{12}}{k} \\ -\frac{a_{21}}{k} & \frac{a_{11}}{k} \end{pmatrix}$$

where $k = a_{11}a_{22} - a_{12}a_{21}$

k is called the **discriminant** of the matrix

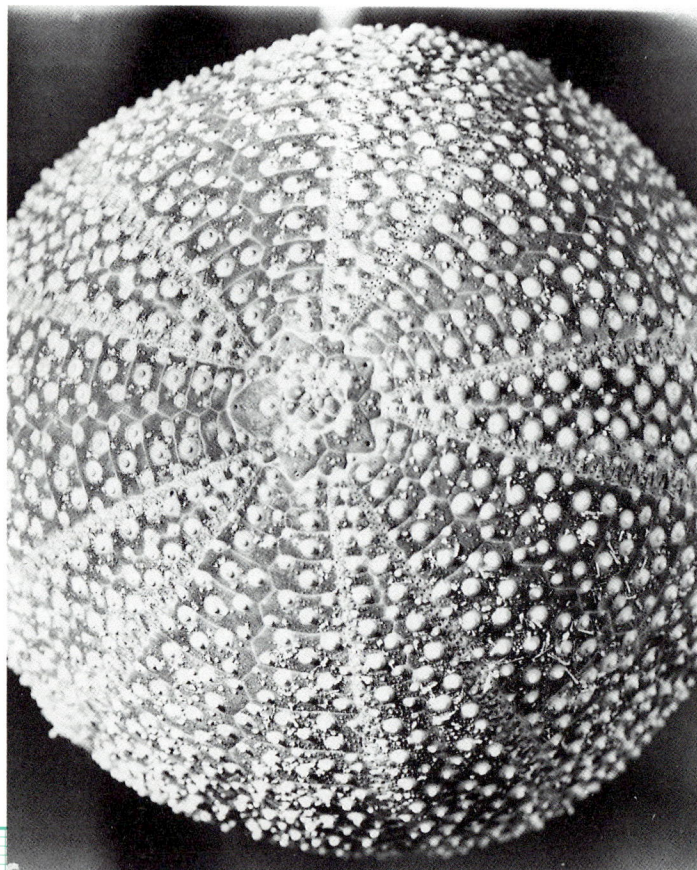

E.g. $\mathbf{A} = \begin{pmatrix} 7 & 3 \\ 4 & 2 \end{pmatrix}$ $\qquad k = 7 \times 2 - 3 \times 4 = 2$

10

$$\mathbf{A}^{-1} = \begin{pmatrix} \frac{2}{2} & -\frac{3}{2} \\ -\frac{4}{2} & \frac{7}{2} \end{pmatrix}$$

$$= \begin{pmatrix} 1 & -\frac{3}{2} \\ -2 & \frac{7}{2} \end{pmatrix}$$

$$\mathbf{A}^{-1}\mathbf{A} = \begin{pmatrix} 1 & -\frac{3}{2} \\ -2 & \frac{7}{2} \end{pmatrix} \begin{pmatrix} 7 & 3 \\ 4 & 2 \end{pmatrix} = \begin{pmatrix} 1 \times 7 + (-\frac{3}{2}) \times 4 & 1 \times 3 + (-\frac{3}{2}) \times 2 \\ (-2) \times 7 + \frac{7}{2} \times 4 & (-2) \times 3 + \frac{7}{2} \times 2 \end{pmatrix}$$

$$= \begin{pmatrix} 1 & 0 \\ 0 & 1 \end{pmatrix}$$

Let's obtain the matrix equations for another transformation
rotation about O through 180° (a half-turn)

$x' = -x$

$y' = -y$

$x' = -1x + 0y$

$y' = 0x + -1y$

$$\begin{pmatrix} x' \\ y' \end{pmatrix} = \begin{pmatrix} -1 & 0 \\ 0 & -1 \end{pmatrix} \begin{pmatrix} x \\ y \end{pmatrix}$$

Try this

12 Obtain the matrix equations for the following transformations

(a) rotation about O through 90° (a quarter-turn anticlockwise)
(b) rotation about O through −90° (a quarter-turn clockwise)

Let's go back to *rotation about O through 180° (a half-turn).*

The transformation matrix is $\begin{pmatrix} -1 & 0 \\ 0 & -1 \end{pmatrix}$.

What is its inverse matrix?

$k = (-1) \times (-1) - 0 \times 0 = 1$

The inverse matrix is $\begin{pmatrix} -1 & 0 \\ 0 & -1 \end{pmatrix}$,
which is the same as the original transformation matrix.

This suggests that the inverse transformation is the same as
the original transformation, i.e. the inverse of *rotation about
O through 180° (a half-turn)* is *rotation about O through 180°
(a half-turn)*

Common sense tells us that this is true — the matrix
method works!

13 (a) Obtain the inverse matrix for the transformation
rotation about O through 90° (quarter-turn anticlockwise)
This inverse matrix is the matrix of which other transformation?
Is this result what you would have expected?

(b) Repeat (a) for the transformation
rotation about O through −90° (quarter-turn clockwise)

Combinations of transformations

Let's consider
a half-turn about O followed by a quarter-turn clockwise about O

You can probably picture the result — the diagram may help you.

$P \to P' \to P''$
A half-turn followed by a quarter-turn clockwise has the same effect as a quarter-turn anticlockwise.

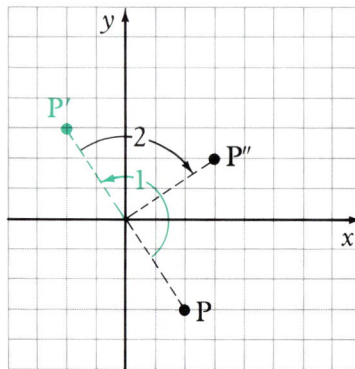

How can we express this double transformation in terms of matrices?

The matrix for a half-turn is $\begin{pmatrix} -1 & 0 \\ 0 & -1 \end{pmatrix}$

The matrix for a quarter-turn clockwise is $\begin{pmatrix} 0 & 1 \\ -1 & 0 \end{pmatrix}$

Let's apply these transformations to the point $(2, -3)$ using matrices.

half-turn on (2, −3): $\quad \begin{pmatrix} -1 & 0 \\ 0 & -1 \end{pmatrix} \begin{pmatrix} 2 \\ -3 \end{pmatrix} = \begin{pmatrix} -2 \\ 3 \end{pmatrix} \quad (2, -3) \longrightarrow (-2, 3)$

quarter-turn on (−2,3): $\quad \begin{pmatrix} 0 & 1 \\ -1 & 0 \end{pmatrix} \begin{pmatrix} -2 \\ 3 \end{pmatrix} = \begin{pmatrix} 3 \\ 2 \end{pmatrix} \quad (-2, 3) \longrightarrow (3, 2)$

If
$\mathbf{M_1}$ is the matrix for the first transformation, T_1,
$\mathbf{M_2}$ is the matrix for the second transformation, T_2 and
(x, y) is the point being transformed,
we can obtain the image point (x'', y'') after the double transformation as follows

$$\begin{pmatrix} x'' \\ y'' \end{pmatrix} = \mathbf{M_2 M_1} \begin{pmatrix} x \\ y \end{pmatrix}$$

The matrix for the double transformation T_1 followed by T_2 is the product of the two matrices $\mathbf{M_2 M_1}$.

Notice the order of the matrices in the product $\mathbf{M_2 M_1}$.

10

For *a half-turn followed by a quarter-turn clockwise*
the product of the two matrices is

$$\begin{pmatrix} 0 & 1 \\ -1 & 0 \end{pmatrix} \begin{pmatrix} -1 & 0 \\ 0 & -1 \end{pmatrix} = \begin{pmatrix} 0 & -1 \\ 1 & 0 \end{pmatrix}$$ the matrix for *a quarter-turn anticlockwise*

a half-turn followed by a quarter-turn clockwise
has the same effect as
a quarter-turn anticlockwise

Applying this product matrix to the point $(2, -3)$ gives

$$\begin{pmatrix} 0 & -1 \\ 1 & 0 \end{pmatrix} \begin{pmatrix} 2 \\ -3 \end{pmatrix} = \begin{pmatrix} 3 \\ 2 \end{pmatrix}$$

which is the result we obtained earlier.

Let's list the transformation matrices we have obtained so far.

TRANSFORMATION	MATRIX
reflection in the line $y = x$	$\begin{pmatrix} 0 & 1 \\ 1 & 0 \end{pmatrix}$
reflection in the line $y = -x$	$\begin{pmatrix} 0 & -1 \\ -1 & 0 \end{pmatrix}$
reflection in the x-axis	$\begin{pmatrix} 1 & 0 \\ 0 & -1 \end{pmatrix}$
reflection in the y-axis	$\begin{pmatrix} -1 & 0 \\ 0 & 1 \end{pmatrix}$
rotation about O through 180° (half-turn)	$\begin{pmatrix} -1 & 0 \\ 0 & -1 \end{pmatrix}$
reflection about O through 90° (quarter-turn anticlockwise)	$\begin{pmatrix} 0 & -1 \\ 1 & 0 \end{pmatrix}$
rotation about O through $-90°$ (quarter-turn clockwise)	$\begin{pmatrix} 0 & 1 \\ -1 & 0 \end{pmatrix}$

Try this

14 In each of the following obtain the double transformation matrix and state which single transformation has the same effect as the double.

(a) a half-turn followed by a quarter-turn anticlockwise
(b) a quarter-turn anticlockwise followed by a quarter-turn clockwise
(c) reflection in the line $y = x$ followed by reflection in the line $y = -x$
(d) reflection in the x-axis followed by reflection in the y-axis
(e) reflection in the line $y = x$ followed by reflection in the x-axis
(f) reflection in the line $y = x$ followed by reflection in y-axis
(g) reflection in the line $y = -x$ followed by reflection in the x-axis
(h) reflection in the line $y = -x$ followed by reflection in the y-axis

Now try this...

A In a unit called Translations in the *Red Book*, the *Now Try This* section posed the question

"*Can you see the connection between* **reflection in two parallel lines** *and* **translation?**"

At that time, we expected that you would guess the connection by looking at a few examples of the double reflection.

Now, however, you have the means to *prove* the result — can you do it?

You might start in this way.

$$P(x,y) \xrightarrow{\text{reflection in } x=k} P'(x',y') \xrightarrow{\text{reflection in } x=j} P''(x'',y'')$$

$x' = 2k - x$
$y' = y$

. . . .

B The diagram shows a rectangle which has undergone a transformation called a **shear**.

A shear is a translation in which movement of each point is
(i) parallel to one axis (the *x*-axis in this case)

and
(ii) proportional to the distance of the point from the axis.

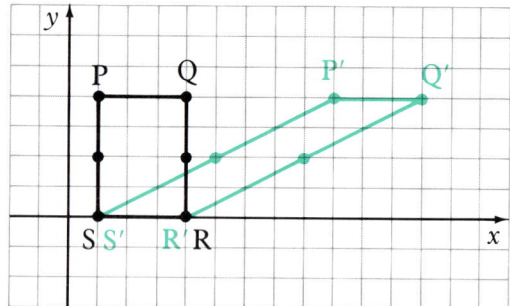

Can you work out the transformation equations?

Here is another example of a shear.

Can you work out the transformation equations for this one?

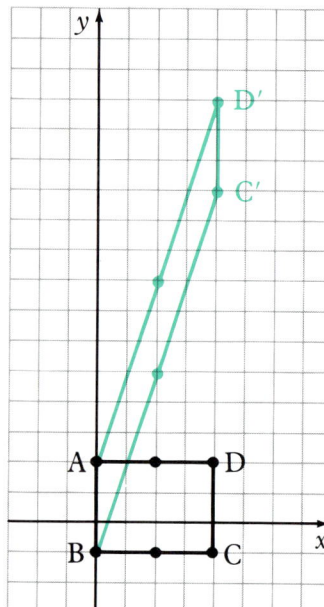

KEY QUESTIONS

10

Please refer to the list of transformation equations and matrices earlier in the unit.

K1 Consider a triangle with vertices P(4,1), Q(2, − 3), R(− 5,0)

Write down the coordinates of the vertices of the image of triangle PQR after
(a) reflection in the x-axis
(b) reflection in $y = -x$

Which single transformation has the same effect as reflection in the x-axis followed by reflection in $y = -x$?

K2 Work out the matrix of the single transformation that has the same effect as *a quarter-turn clockwise* followed by *a half-turn*.

Name the single transformation.

STATISTICS

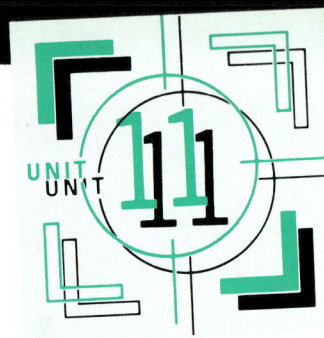

Two small companies pay salaries as follows.

PATERNAL PLASTICS EMPLOYEES
9 are paid £9000 each
9 are paid £11000 each
2 are paid £25000 each

EXPLOIT ELECTRONICS EMPLOYEES
18 are paid £8000 each
1 is paid £26000
1 is paid £60000

The 18 lower-paid employees of Exploit Electronics claim that ". . . salaries in Paternal Plastics are higher than ours . . ." only to be told, by their Personnel Manager (who earns £26000), that ". . . the average salary in both companies is the same . . ."

The angry 18 are not satisfied — would you be?

What is meant by *average*?
What is the most effective way of comparing salaries in the two companies?

These are questions which we will try to answer in this unit.

Remember, as someone once said, ". . . there are lies, damned lies and statistics . ."

You may recall a unit in a previous book ('Graphs and Charts' in the *Yellow Book*) where you were asked to put the following weights (in kg) of 20 students into a stem-leaf table.

47	52	62	37	49
39	38	46	60	52
62	43	54	44	48
54	55	51	56	55

The stem-leaf table (with the smallest numbers first) looks like this

```
3 | 7 8 9      (37, 38, 39)
4 | 3 4 6 7 8 9
5 | 1 2 2 4 4 5 5 6
6 | 0 2 2
```

Try these

1 Here are the heights (in cm) for the same class of students.

```
169  173  179  171  183
181  185  167  180  175
174  169  182  165  181
181  171  180  173  169
```

Make a stem-leaf table for the heights.

2 Class X and class Y each has twenty students. The students were asked to donate all their loose change to charity.
Here are the results (amounts in pence).

```
X   79  67  58  25  45
    49  38  75  65  55
    87  29  12  91  83
    57  68  78  99  82

Y   34  71  59  62  70
    12  91  66  63  44
    80  81  90  48  13
    74  53  82  70  72
```

Make a stem-leaf table for each of the classes.

Here are two stem-leaf tables showing the marks of two classes of 30 students in an examination.

CLASS A

```
4 | 7 7 7 8 8 8 9 9
5 | 2 2 2 2 6 7 7 9 9 9
6 | 1 1 3 3 5 7 7 8
7 | 2 3 7
8 | 1
```

CLASS B

```
3 | 8 9 9
4 | 3 5 5 7 7 8
5 | 0 2 4 4 4 6 7
6 | 5 7 7 9
7 | 3 5 5 6 6
8 | 0 1 2
9 | 0 1
```

Someone has asked..."Which class had the better examination results?" Let's examine both sets of marks to try to find an answer to the question.

The first statistic that we might look at is the **average** mark. There is more than one kind of average.

The first kind that we will consider is the **mean**. The **mean mark** is calculated by adding the marks for the class and dividing by the number of students.

For class A, the mean mark is

$$\frac{1756}{30} \text{ (total of 30 marks)}$$

$= 58\cdot5$ (to 1 decimal place)

Try these

3 Calculate the mean mark for class B.

4 Calculate the mean amounts for class X and class Y in Try This 2.

If all that is required is an approximate mean, it can be calculated more quickly as follows

CLASS A

```
4 | 7 7 7 8 8 8 9 9
5 | 2 2 2 6 7 7 9 9 9
6 | 1 1 3 3 5 7 7 8
7 | 2 3 7
8 | 1
```

Assume that all marks in the 40's are the same and choose the mid-point of the 40's to be that mark.
The mid-point is $44\cdot5$ as we can see from this number line.

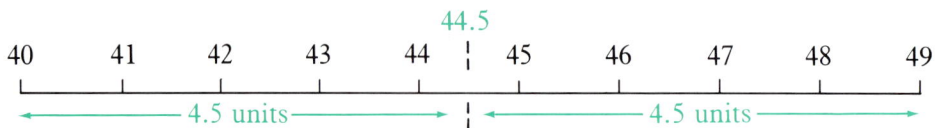

We have   8 marks of $44\cdot5$ giving a total of $8 \times 44\cdot5 = 356$ for the 40's
and   10 marks of $54\cdot5$ giving a total of $10 \times 54\cdot5 = 545$ for the 50's
          8 marks of $64\cdot5$ giving a total of $8 \times 64\cdot5 = 516$ for the 60's
          3 marks of $74\cdot5$ giving a total of $3 \times 74\cdot5 = 223\cdot5$ for the 70's
          1 marks of $84\cdot5$ giving a total of $1 \times 84\cdot5 = 84\cdot5$ for the 80's

We have 30 marks....giving a total of 1725

giving a mean mark of $\dfrac{1725}{30}$

$= 57$ or 58 (to the nearest 1)
(compared with $58\cdot5$ by the accurate method)

Let's set this down in a **grouped frequency** table.

11

MARK	MID-POINT (M)	FREQUENCY (f)	f × M
40–49	44·5	8	356
50–59	54·5	10	545
60–69	64·5	8	516
70–79	74·5	3	223·5
80–89	84·5	1	84·5
	TOTALS	30	1725

(The marks are grouped in *intervals* 40–49, 50–59, etc.)

$$\text{Mean} = \frac{1725}{30} = 57 \text{ or } 58 \quad \text{(to nearest 1)}$$

Try these

5 Complete a table like the one above for the marks of class B and calculate the approximate mean.
(Use the same intervals as before, i.e. 40–49, 50–59, etc.)

6 Complete tables and calculate approximate means for the amounts donated by each class in Try This 2.
(Use intervals 0–19, 20–39, 40–59, etc.)

Let's now look at another kind of average, **the median**.

The **median mark** is the one that is 'in the middle' after all the marks have been put in order from lowest to highest.

For 30 marks, the median is 'between' the 15th and 16th mark.
The stem-leaf table helps us to identify the 15th and 16th marks easily.

CLASS A

```
     1st ————————→ 8th
4 |  7 7 7 8 8 8 9 9

     9th ————————→
5 |  2 2 2 2 6 7 7 9 9 9      57 is the 15th mark
6 |  1 1 3 3 5 7 7 8          59 is the 16th mark
7 |  2 3 7
8 |  1
```

The median is half-way between 57 and 59, i.e. at 58.

Note: (i) Had the number of marks been 31, the median would have been the 16th mark. We had to find half-way between the 15th and 16th marks because there were an even number of marks (30).

(ii) In this case, the median and the mean are approximately the same.

7 Find the median of the marks for class B.

8 Find the median for the amounts donated by each class in Try This 2.

It may be useful to know, for example, *"How many students scored less than 70 in the examination?"*

To obtain this statistic, we look at **cumulative frequency**.

Let's do this for the marks of class A by extending the frequency table that we obtained previously.

MARK	MID-POINT	FREQUENCY	CUMULATIVE FREQUENCY	
40–49	44·5	8	8	
50–59	54·5	10	18	(8+10=18)
60–69	64·5	8	26	(18+8=26)
70–79	74·5	3	29	(26+3=29)
80–89	84·5	1	30	(29+1=30)

We can interpret the cumulative frequencies as follows:

"8 students scored 49 marks or less"
            or "8 students scored less than 50"

"18 students scored 59 marks or less"
            or "18 students scored less than 60"

"26 students scored 69 marks or less"
            or "26 students scored less than 70"

"29 students scored 79 marks or less"
            or "29 students scored less than 80"

"30 students scored 89 marks or less"
            or "30 students scored less than 90"

A more useful form for the cumulative frequency table is the following.

MARK	LESS THAN	FREQUENCY	CUMULATIVE FREQUENCY
40–49	50	8	8
50–59	60	10	18
60–69	70	8	26
70–79	80	3	29
80–89	90	1	30

11

We can now draw a frequency graph and a cumulative frequency graph.

The frequency graph can be a bar graph or a line graph depending on whether we use intervals or mid-points along the Marks axis.

The cumulative frequency graph is a line graph using 'less than' along the Marks axis.

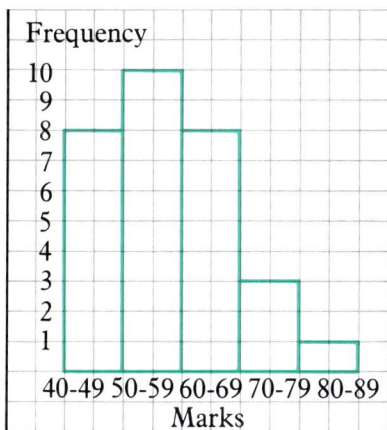

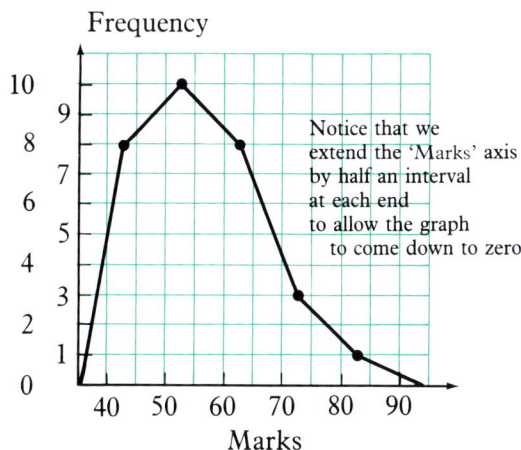

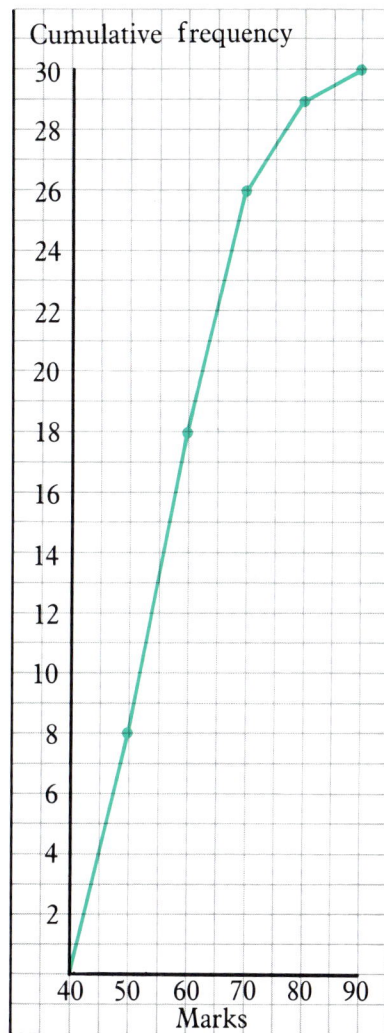

Notice that we extend the 'Marks' axis by half an interval at each end to allow the graph to come down to zero

Try these

9 Construct a cumulative frequency table, and draw frequency and cumulative frequency line graphs for the marks of class B.
(Use the frequency table that you obtained in Try This 5.)

10 Construct a cumulative frequency table, and draw frequency and cumulative frequency line graphs for each of the frequency tables that you obtained in Try This 6.

We may be interested in the extent to which marks are spread around the mean. One such measure of spread or **dispersion** is the **range**.

The range of marks for class A is
 highest mark − lowest mark
= 81 − 47
= 34

As we have noted, the median is the mark 'half-way up' the order (in this case between the 15th and 16th marks).

The mark that is 'quarter-way up' the order is called the lower **quartile**.
In this case, the lower quartile is the 8th mark.

```
4 | 7 7 7 8 8 8 9 9
5 | 2 2 2 2 6 7 7 9 9 9
6 | 1 1 3 3 5 7 7 8
7 | 2 3 7
8 | 1
```

The 8th mark is 49. The lower quartile $Q_1 = 49$.

Similarly, the 23rd mark is 65. The upper quartile $Q_3 = 65$.

The middle quartile, Q_2, is the median.

The difference   upper quartile − lower quartile
is called the **interquartile range** and is another measure of spread.

For class A, the interquartile range is $65 − 49 = 16$.

It is more common to use the **semi-interquartile range**, which is $\frac{1}{2}$(upper quartile − lower quartile).

For class A, the semi-interquartile range is 8.

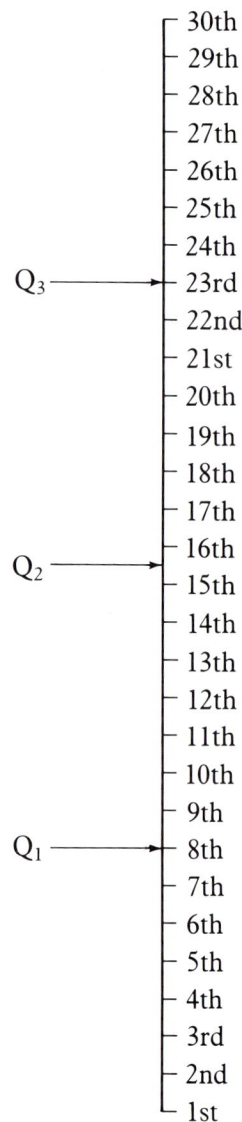

$Q_3 \longrightarrow$ 23rd

$Q_2 \longrightarrow$ 15th

$Q_1 \longrightarrow$ 8th

30th
29th
28th
27th
26th
25th
24th
23rd
22nd
21st
20th
19th
18th
17th
16th
15th
14th
13th
12th
11th
10th
9th
8th
7th
6th
5th
4th
3rd
2nd
1st

Try these

11 Find the quartiles and the interquartile range for class B.

12 Find the quartiles and the semi-interquartile range for each of the two data sets in Try This 2 (the money donated by classes X and Y).

You can see how the quartiles appear on the cumulative frequency graph of the marks of class A.

We have 'smoothed out' the cumulative frequency **polygon** that we drew earlier and obtained a cumulative frequency **curve**.

The quartiles Q_1 and Q_3 and the median Q_2 can be obtained approximately from the cumulative frequency curve, which shows that
Q_1 is about 50
Q_2 is about 57
Q_3 is about 66

You could try this for one of the cumulative frequency graphs that you have already drawn.

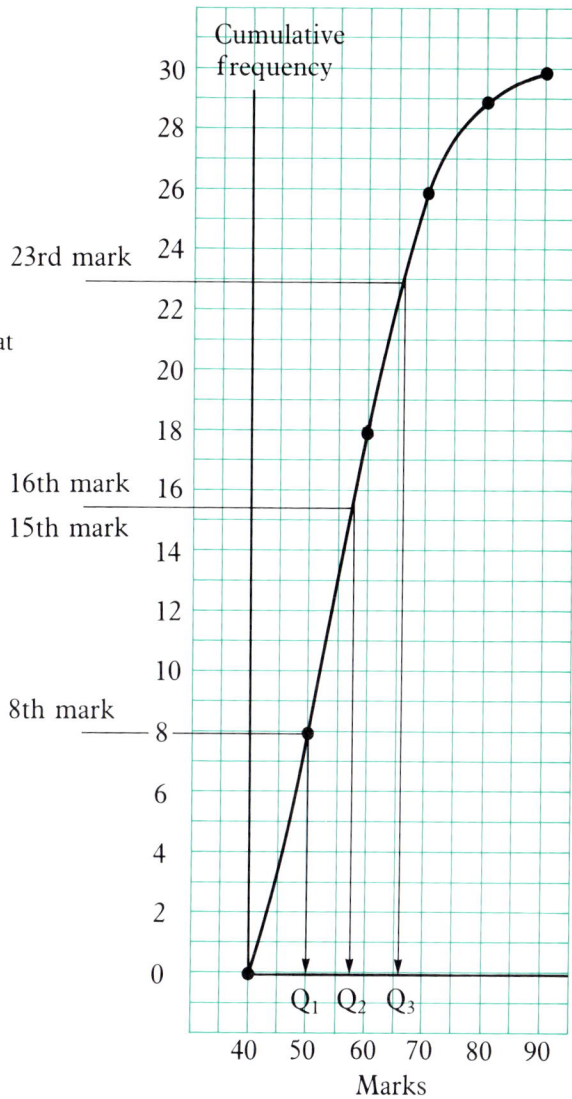

11

The third kind of average that we will consider is the **mode**.
The mode is the mark that appears most frequently.

For class A, the mode is 52.

```
4 | 7 7 7 8 8 8 9 9
5 | 2 2 2 2 6 7 7 9 9 9
6 | 1 1 3 3 5 7 7 8
7 | 2 3 7
8 | 1
```

If the data have been grouped into intervals as below, we talk about the **modal interval** or **modal class** rather than the mode.

MARK	MID-POINT (M)	FREQUENCY (f)	
40–49	44·5	8	
50–59	54·5	**10**	50–59 is the modal class
60–69	64·5	8	
70–79	74·5	3	
80–89	84·5	1	

Try this

13 (a) Write down the mode for the marks of class B.

(b) Assuming that the marks for class B have been grouped in classes (40–49, etc), write down the modal class.

Let's summarise the statistics that we have obtained for the marks of classes A and B.

	mean	median	mode	range	semi-interquartile range
CLASS A	58·5	58	52	34	8
CLASS B	61·2	56·5	54	53	14

Class B has the higher average mark if we choose to look at the mean, whereas class A has the higher average mark if we choose to look at the median.

The marks of class A are confined between 47 and 81 whereas the marks of class B range between 38 and 91.

What all this shows is that it is important, when we use statistics, to know why we are doing so and to choose the statistics that most suit our purposes.

It may be, for example, that classes A and B are taught in different ways. Perhaps students in class B, where the marks are more widely ranged, are free to work at their own pace and get ahead of (or fall behind!) their classmates. Class A may be taught in a more controlled way which keeps marks closer together. Those who prefer one of the two possible ways of teaching to the other might, somewhat dishonestly, use the statistics that most strongly support their arguments!

Now try this... **11**

A We used two separate stem-leaf tables to show the marks of classes A and B. Combine the two tables in a way that allows comparison to be made as effectively as possible.

B We looked at the extent to which marks were spread out by calculating the range and the semi-interquartile range.

Another way of looking at spread (or **dispersion**) is to find the difference between each mark and the mean mark and then to calculate the mean of all the differences. This is called the **mean deviation** of the data set.

As we have defined it so far, mean deviation will be zero! How do we get round this problem?
What is the mean deviation for the marks of class A?
How does it compare with the semi-interquartile range for these marks?

There is yet another measure of spread called **standard deviation** which you may wish to investigate and compare with other measures of dispersion.

KEY QUESTION

K1

113	81	91	117	107
107	84	103	97	89
117	85	113	84	101
98	95	105	105	115
108	93	103	104	115

For the above data set,

(a) draw a stem-leaf table;

(b) calculate the exact mean;

(c) construct a grouped frequency table with intervals 80–89, etc and use it to calculate an approximate mean;

(d) extend the table in (c) to include cumulative frequency;

(e) write down the median, the upper and lower quartiles and the semi-interquartile range;

(f) draw frequency and cumulative frequency line graphs.

REMINDERS AND REVISION — FACTORS

REMINDER EXAMPLE 1

Factorise $6x^2y + 9y^2z$

$$6x^2y + 9y^2z = 2 \times 3 \times x \times x \times y + 3 \times 3 \times y \times y \times z$$
$$= 3y(2x^2 + 3yz)$$

REMINDER EXAMPLE 2

Factorise $x^2 - 2x - 15$

$$x^2 - 2x - 15 = (x + N_1)(x + N_2)$$
$$= x^2 + (N_1 + N_2)x + N_1N_2$$

$$\text{add to } -2 \quad \text{multiply to } -15$$

$$x^2 - 2x - 15 = (x - 5)(x + 3)$$

(after probably more than one try.)

REMINDER EXAMPLE 3

Factorise $6y^2 - 23yz + 20z^2$

$$6y^2 - 23yz + 20z^2 = (M_1y + N_1z)(M_2y + N_2z)$$
$$= M_1M_2y + (N_1M_2 + N_2M_1)yz + N_1N_2z$$

$$\text{multiply to } 6 \quad -23 \quad \text{multiply to } 20$$

$$6y^2 - 23yz + 20z^2 = (2y - 5z)(3y - 4z)$$

(after more than one try.)

REMINDER EXAMPLE 4

Factorise $25a^2 - 9b^2$

$$25a^2 - 9b^2 = (5a)^2 - (3b)^2$$
$$= (5a - 3b)(5a + 3b)$$

REMINDER EXAMPLE 5

Simplify $\dfrac{2 - 8p^2}{2 - p - 6p^2}$

$$2 - 8p^2 = 2(1 - 4p^2) = 2(1 - 2p)(1 + 2p)$$
$$2 - p - 6p^2 = (1 - 2p)(2 + 3p)$$
$$\frac{2 - 8p^2}{2 - p - 6p^2} = \frac{2(1 - 2p)(1 + 2p)}{(1 - 2p)(2 + 3p)}$$
$$= \frac{2(1 + 2p)}{2 + 3p}$$

REVISION EXERCISE

1 Multiply

(a) $(x+2)(x+3)$ (b) $(2p-q)(5p-q)$

(c) $(3m-2n)(3n-2m)$ (d) $(4r-3s)(2r+5s)$

2 Factorise completely

(a) $8m^2n-12mn^2$

(b) $x^2+10x+21$

(c) $2p^2+4pq-6q^2$

(d) $36y^2-49z^2$

(e) $d^3e^2f+e^3f^2d-f^3d^2e$

(f) $2-5r+2r^2$

(g) $12a^2-ab-6b^2$

(h) $27-12c^2$

(i) $2x^2+2x+4$

(j) $30s^2+25st-105t^2$

(k) $12c^2-31cd+20d^2$

(l) $1-100z^2$

3 Simplify each of these expressions

(a) $\dfrac{2x-2}{x^2-1}$ (b) $\dfrac{x^2-xy-6y^2}{x^2-5xy+6y^2}$

(c) $\dfrac{8p^2-18}{6p^2-5p-6}$ (d) $\dfrac{2-2a}{2-2a^2}$

4 A square has side L units. A rectangle has the same length as the square, but is 2 units less in width.
What is the difference between the areas of the square and the rectangle?
Illustrate your answer with a diagram.

5 A rectangle is 2 units more in length and 2 units less in width than a square.
What is the difference between the areas of the square and the rectangle?
Illustrate your answer with a diagram.

6 A rectangle is 2 units more in length and 3 units less in width than a square.
What is the difference between the areas of the square and the rectangle?
Illustrate your answer with a diagram.

REMINDERS AND REVISION — STRAIGHT LINE AND INEQUALITIES

Reminder facts

EQUATION OF A STRAIGHT LINE

When the equation of a line is stated in the form

$y = ax + b$

a is the gradient of the line and b is the intercept on the y-axis.

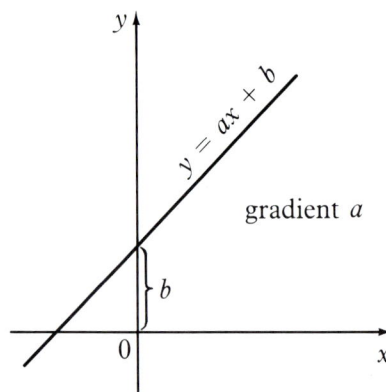

INEQUATIONS

Inequations can be stated in various forms such as $x + 2y \leqslant 6$.

$\leqslant$ means less than or equal to.

The solution set $x + 2y \leqslant 6$ is shown by the solid line and the shading.

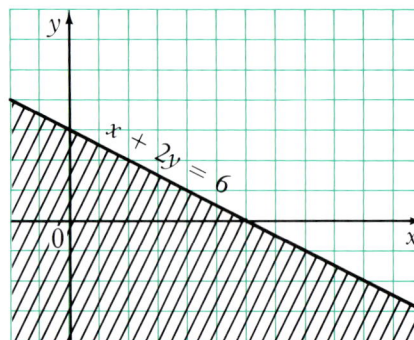

The solution set of $2x + y > 8$ is shown opposite.

Note the use of a broken line.

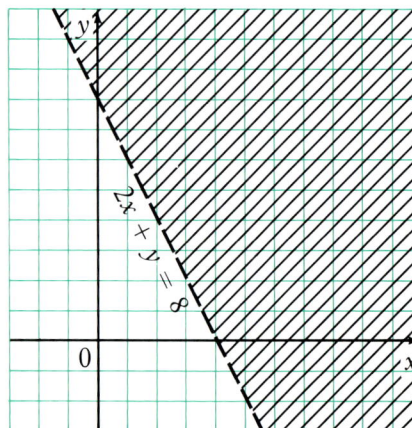

INEQUALITIES

The same rules for solving equations are used when solving inequalities, *but* when dividing or multiply both sides by a negative number the symbol is reversed.

REMINDER EXAMPLE

Solve $2(x-3)-3(2x+1) \leqslant 3$

$$2x-6-6x-3 \leqslant 3$$
$$-4x-9 \leqslant 3$$
$$-4x \leqslant 3+9$$
$$-4x \leqslant 12$$
$$x \geqslant -3$$

REVISION EXERCISE

1 State the gradient and the intercept on the y-axis of the following lines.

(a) $y = 2x - 3$ (d) $x + 2y = 6$

(b) $y = 4 - 3x$ (e) $y - 3x - 5 = 0$

(c) $2x + y = 0$ (f) $\frac{1}{3}x + \frac{1}{2}y = -2$

2 Find the equations of the lines whose gradients and intercepts on the y-axis are given:

(a) gradient = 3, intercept = 4;

(b) gradient = -2, intercept = 0;

(c) gradient = $\frac{1}{2}$, intercept = -2.

3 Find the equation of the line whose intercept on the y-axis is -2 and which is parallel to the line $y - 2x = 4$.

4 Find the equation of the line that passes through the following pairs of coordinates.

(a) $(0,0)$, $(5,5)$

(b) $(3,1)$, $(6,4)$

(c) $(-2,-2)$, $(2,0)$

(d) $(-1,3)$, $(1,-1)$

5 Describe the areas that are shaded.

6 Draw the line with equation $y = 4 - 2x$. Show the solution set for $y \geqslant 4 - 2x$.

7 Draw the line with equation $2x + 3y - 6 = 0$. Show the solution set for $2x + 3y - 6 < 0$.

8 Solve for x.

(a) $x + 3 > 5$

(b) $3x + 1 \leqslant x + 7$

(c) $5x - 3 \geqslant 15 - x$

(d) $x - 7 \leqslant 23 + 6x$

(e) $3(2x - 1) - 3(4x + 5) \leqslant 0$

(a)

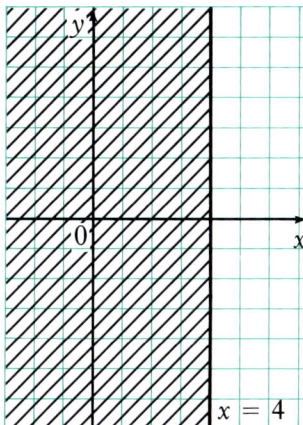

$x = 4$

(b)

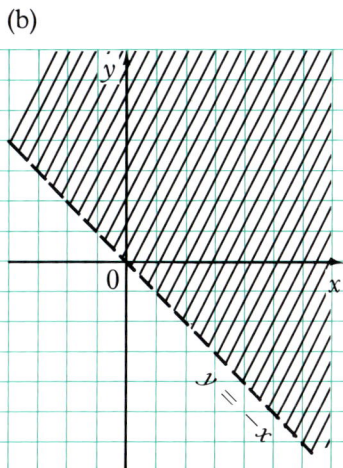

(c)

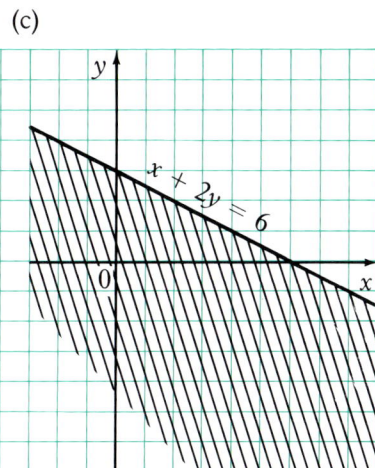

$x + 2y = 6$

REMINDERS AND REVISION — SIMULTANEOUS EQUATIONS AND LINEAR PROGRAMMING

Reminder facts

SIMULTANEOUS EQUATIONS

A pair of simultaneous equations with two unknowns must be so altered that by adding or subtracting the equations one of the unknowns is eliminated.

REMINDER EXAMPLE

Solve: (1) $2a + 3b = 1$
　　　　(2) $3a - 4b = 10$

Multiply equation (1) by 3 and equation (2) by 2.
$6a + 9b = 3$
$6a - 8b = 20$

Subtract the equations to eliminate a.
$(6a + 9b) - (6a - 8b) = 3 - 20$
$6a + 9b - 6a + 8b = -17$
$17b = -17$
$b = -1$

Substitute $b = -1$ in equation (1) or (2).
(1) $2a + 3b = 1$
　　　$2a - 3 = 1$
　　　　$2a = 4$
　　　　　$a = 2$

$a = 2$
$b = -1$

Check that your answer is correct by substituting your solutions into equation (2).

LINEAR PROGRAMMING

In the diagram opposite reverse shading has been used (i.e. the area not involved has been scored out).

The clear region is defined by:
$$x \geqslant 0$$
$$y \geqslant 0$$
$$2x + y \leqslant 8$$
$$x + 2y \leqslant 6$$

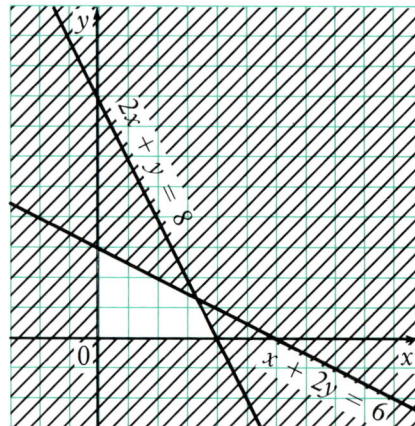

MAXIMUM AND MINIMUM VALUES

To find the maximum or minimum values either check the
value of individual points or draw a searchline.
Remember: in this unit we use only integer values.

REMINDER EXAMPLE

To find the maximum value of $x+y$ for the inequations
shown above, either check individual points:

minimum value of $x+y$ is 0 at $(0,0)$
maximum value of $x+y$ is 4 at $(2,2)$, $(3,1)$ or $(4,0)$,

or draw searchlines: $x+y=k$ where $k=0$, 1, 2 etc.

REVISION EXERCISE

1 Solve the following simultaneous equations:

(a) $x+y=5$
$x-y=1$

(b) $2x+3y=1$
$2x-y=5$

(c) $2a-3b=-13$
$a+2b=4$

(d) $-3x-4y=11$
$2x+3y=-8$

(e) $2a+4b=-7$
$3a-5b=17$

(f) $2x-3y-5=0$
$2y+5x+16=0$

2 (a) Describe the clear region using
inequations.

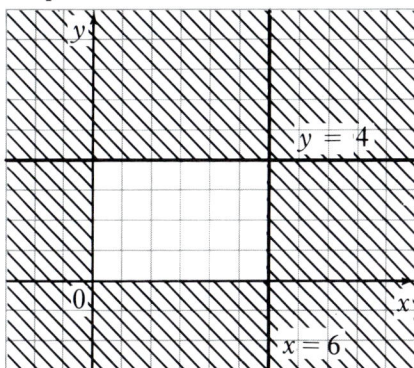

(b) Find the maximum and minimum
value of $x+y$ in this region.

3 (a) Describe the clear region using
inequations.

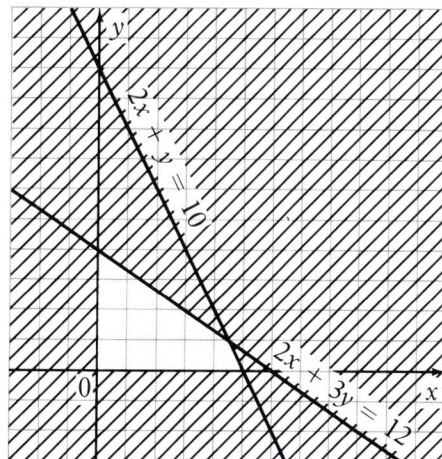

(b) Find the maximum and minimum value
of $2x+y$ in this region.

4 On separate diagrams, illustrate the regions
defined by the following sets of inequations:

(a) $x\geqslant0$, $y\geqslant0$, $x\leqslant5$, $y\leqslant6$
(b) $x\geqslant-1$, $y\geqslant2$, $x\leqslant6$, $y\leqslant6$
(c) $x\geqslant0$, $y\leqslant0$, $2y-x\leqslant6$, $2x+y\leqslant10$
(d) $x\geqslant0$, $y\leqslant0$, $x+2y\leqslant8$, $y\leqslant2x+2$

REMINDERS AND REVISION —
INDICES AND ALGEBRAIC FRACTIONS

Reminder facts

INDICES

$$a^m \times a^n = a^{m+n}$$
$$a^m \div a^n = a^{m-n}$$
$$a^0 = 1$$
$$a^{-m} = \frac{1}{a^m}$$
$$(a^m)^n = a^{mn}$$
$$a^{\frac{m}{n}} = \sqrt[n]{a^m} \quad \text{or} \quad (\sqrt[n]{a})^m$$

Surds
$$\sqrt{ab} = \sqrt{a} \times \sqrt{b}$$
$$\sqrt{\frac{a}{b}} = \frac{\sqrt{a}}{\sqrt{b}}$$

REMINDER EXAMPLES

$$x^3 \times x^2 = x^5$$
$$x^6 \div x^4 = x^2$$
$$5^0 = 1$$
$$x^{-2} = \frac{1}{x^2}$$
$$(x^3)^2 = x^6$$
$$9^{\frac{3}{2}} = (\sqrt{9})^3 = 3^3 = 27$$

$$\sqrt{4 \times 9} = \sqrt{4} \times \sqrt{9}$$
$$\sqrt{\frac{16}{4}} = \frac{\sqrt{16}}{\sqrt{4}}$$

ALGEBRAIC FRACTIONS

$$\frac{x}{y} \times \frac{p}{q} = \frac{xp}{yq}$$

$$\frac{x}{y} \div \frac{p}{q} = \frac{x}{y} \times \frac{q}{p} = \frac{xq}{yp}$$

$$\frac{x}{y} + \frac{p}{q} = \frac{xq}{yq} + \frac{py}{qy} = \frac{xq + py}{yq}$$

$$\frac{x}{y} - \frac{p}{q} = \frac{xq}{yq} - \frac{py}{qy} = \frac{xq - py}{yq}$$

REMINDER EXAMPLE

Solve this equation.
$$\frac{x}{5} = \frac{x-4}{4}$$

Common denominator is 20
$$\frac{4x}{20} = \frac{5(x-4)}{20}$$

Equate numerators
$$4x = 5(x-4)$$
$$4x = 5x - 20$$
$$4x - 5x = -20$$
$$x = 20$$

REVISION EXERCISE

1 Simplify (leaving answer in index form)

(a) $p^3 \times p^4$

(b) $2x^5 \div 3x^2$

(c) $3a^2 \times 4a^3$

(d) $6b^7 \div 2b^5$

(e) $\dfrac{2^4 \times 2^3}{2^2}$

(f) $\dfrac{10x^2}{5x^3}$

2 Evaluate

(a) 3^{-2}

(b) $7^3 \times 7^{-3}$

(c) $(2^{\frac{1}{2}})^2$

(d) $5^2 \div 5^0$

(e) $(4^3)^{\frac{1}{3}}$

(f) $36^{-\frac{1}{2}}$

3 Simplify (leaving answers in index form)

(a) $3ab^2 \times 2a^2b$

(b) $a^2(a^{-2} + a^3)$

(c) $\dfrac{5x^2y}{10xy^2}$

(d) $\dfrac{6p^3q}{2pq}$

4 (a) Express 6 million in standard form.
   (b) Express 2·5 billion in standard form.

5 Calculate the following and then express in standard form.

(a) $(8 \times 10^4) \times (2 \times 10^3)$

(b) $\dfrac{2 \times 10^2}{4 \times 10^3}$

6 Simplify the following (leave answers as surds).

(a) $\sqrt{18}$

(b) $\sqrt{300}$

(c) $\sqrt{6} \times \sqrt{12}$

(d) $\sqrt{15} \times \sqrt{5}$

7 Rationalise the denominators of the following fractions and simplify where possible.

(a) $\dfrac{1}{\sqrt{7}}$

(b) $\dfrac{4}{\sqrt{6}}$

(c) $\dfrac{16}{2\sqrt{8}}$

8 Write each of the following as a single fraction.

(a) $\dfrac{2a^2}{b} \times \dfrac{a}{b}$

(b) $\dfrac{a}{2} \div \dfrac{b}{a}$

(c) $b \times \dfrac{a}{b}$

(d) $\dfrac{p^2}{q^2} \div \dfrac{p}{q}$

9 Write each of the following as a single fraction.

(a) $\dfrac{1}{x} + \dfrac{1}{y}$

(b) $\dfrac{a}{2b} - \dfrac{4}{ab}$

(c) $\dfrac{1}{x^2 - 2} - \dfrac{1}{x^2}$

(d) $\dfrac{2}{x+5} - \dfrac{3}{x-5}$

10 Solve each of these equations.

(a) $\dfrac{x+2}{3} = x + 6$

(b) $\dfrac{14}{2x+1} = \dfrac{4}{x-1}$

REMINDERS AND REVISION — QUADRATICS

Consider the quadratic equation
$2x^2 + 5x - 12 = 0 \qquad x \in R$

(a) factorise the quadratic expression
(b) solve the quadratic equation
(c) calculate the minimum value of the expression

(a) $2x^2 + 5x - 12 = (2x - 3)(x + 4)$

(b) $\quad 2x^2 + 5x - 12 = 0$
$\quad (2x - 3)(x + 4) = 0$
$\quad 2x - 3 = 0 \qquad$ and $\qquad x + 4 = 0$
$\qquad x = \frac{3}{2} \,(1 \cdot 5) \qquad\qquad x = -4$

(c) minimum value when x is halfway between roots
$\quad x = \frac{1}{2}[1 \cdot 5 + (-4)]$
$\qquad = -1 \cdot 25$

$\quad$ minimum value $= 2(-1 \cdot 25)^2 + 5(-1 \cdot 25) - 12$
$\qquad\qquad\qquad\quad = 2 \times 1 \cdot 5625 + (-6 \cdot 25) - 12$
$\qquad\qquad\qquad\quad = 3 \cdot 125 - 6 \cdot 25 - 12$
$\qquad\qquad\qquad\quad = -15 \cdot 125$
$\qquad\qquad$ when $x = -1 \cdot 25$

REMINDER EXAMPLE 2

Solve the quadratic equation
$4x(x - 3) = (x - 1)(x + 2) \qquad x \in R$

$4x(x - 3) = (x - 1)(x + 2)$
$4x^2 - 12x = x^2 + x - 2$
$4x^2 - 12x - x^2 - x + 2 = 0$
$3x^2 - 13x + 2 = 0$

$3x^2 - 13x + 2$ does not factorise and we must use the
quadratic formula which states that
the roots of the quadratic equation $ax^2 + bx + c = 0$
are given by

$$x = \frac{-b + \sqrt{(b^2 - 4ac)}}{2a} \qquad \textbf{or} \qquad x = \frac{-b - \sqrt{(b^2 - 4ac)}}{2a}$$

$3x^2 - 13x + 2 = 0$
$a = 3, \ b = -13, \ c = 2$

$$x = \frac{-(-13) + \sqrt{[(-13)^2 - 4 \times 3 \times 2]}}{2 \times 3} \qquad \text{or} \qquad x = \frac{-(-13) - \sqrt{[(-13)^2 - 4 \times 3 \times 2]}}{2 \times 3}$$

$$x = \frac{13 + \sqrt{(169 - 24)}}{6} \qquad \text{or} \qquad x = \frac{13 - \sqrt{(169 - 24)}}{6}$$

$$x = \frac{13 + \sqrt{145}}{6} \qquad \text{or} \qquad x = \frac{13 - \sqrt{145}}{6}$$

$$x = \frac{13 + 12 \cdot 04}{6} \quad \text{(4 sig. figs)} \qquad \text{or} \qquad x = \frac{13 - 12 \cdot 04}{6} \quad \text{(4 sig. figs)}$$

$$x = \frac{25 \cdot 04}{6} \qquad\qquad\qquad \text{or} \qquad x = \frac{0 \cdot 96}{6}$$

$$x = 4 \cdot 17 \quad \text{(3 sig. figs)} \qquad \text{or} \qquad x = 0 \cdot 160 \quad \text{(3 sig. figs)}$$

REMINDER EXAMPLE 3

Solve the equation

$$\frac{1}{x} - \frac{2}{3x - 1} = \frac{1}{12} \qquad x \in R$$

We multiply each term by $x \times (3x - 1) \times 12$

$$\frac{1 \times x \times (3x - 1) \times 12}{x} - \frac{2 \times x \times (3x - 1) \times 12}{(3x - 1)} = \frac{1 \times x \times (3x - 1) \times 12}{12}$$

and cancel within each fraction in the equation

$$\frac{1 \times \cancel{x} \times (3x - 1) \times 12}{\cancel{x}} - \frac{2 \times x \times \cancel{(3x - 1)} \times 12}{\cancel{(3x - 1)}} = \frac{1 \times x \times (3x - 1) \times \cancel{12}}{\cancel{12}}$$

Giving

$$1 \times (3x - 1) \times 12 - 2 \times x \times 12 = 1 \times x \times (3x - 1)$$
$$36x - 12 \quad - \quad 24x \quad = 3x^2 - x$$
$$-3x^2 + 36x - 24x + x - 12 = 0$$
$$-3x^2 + 13x - 12 = 0$$
$$3x^2 - 13x + 12 = 0$$
$$(x - 3)(3x - 4) = 0$$
$$x = 3, \quad x = \frac{4}{3}$$

REVISION EXERCISE

1 In each of the following

factorise the quadratic expression;
solve the quadratic equation on R;
calculate the minimum value of the
expression and state the value of x which
gives this minimum.

(a) $x^2 + 6x + 8 = 0$
(b) $x^2 - 6x + 5 = 0$
(c) $x^2 - 4x - 5 = 0$
(d) $x^2 + 5x - 6 = 0$

2 In each of the following

factorise the quadratic expression;
solve the quadratic equation on R;
calculate the maximum value of the
expression and state the value of x which
gives this maximum.

(a) $28 - 3x - x^2 = 0$
(b) $18 + 3x - x^2 = 0$
(c) $28 - 27x - x^2 = 0$
(d) $18 + 17x - x^2 = 0$

3 In each of the following calculate the
maximum or minimum value of the
expression.

(a) $55 - 4x - 3x^2$
(b) $x^2 + 8x + 12$
(c) $36 + 6x - 2x^2$
(d) $15x^2 + 4x - 4$
(e) $x^2 - 6x$
(f) $2x - x^2$

4 Solve each of the following quadratic equations on R

    (a) $x(x-1)=30$
    (b) $x(x+3)=4$
    (c) $9x-x^2=-10$
    (d) $x(8-x)=12$
    (e) $4x(x-2)=(x-1)(x-5)$
    (f) $(x+1)(x+8)=3x(x-2)$

5 Without using the quadratic formula, find, to 3 significant figures, the roots of the equation
$$x^2-8x=-6 \qquad x\in R$$

    You may draw graphs and you should use a calculator.

6 Use the quadratic formula to solve the following equations on R giving roots to 3 significant figures where appropriate.

    (a) $x^2-11x+16=0$
    (b) $x^2-9x=-5$
    (c) $x^2=7x+2$
    (d) $3-5x-x^2=0$
    (e) $x(x-2)=10$
    (f) $x(x+4)=8$
    (g) $3x-x^2=-6$
    (h) $x(8-x)=11$
    (i) $(x-2)(x-9)=3x(x-4)$
    (j) $4x(x+3)=(x+3)(x-4)$

7 Solve each of the following equations on R, giving the roots to 3 significant figures where necessary.

    (a) $\dfrac{1}{x+2}+\dfrac{1}{x+3}=\dfrac{7}{12}$

    (b) $\dfrac{x+1}{2}-\dfrac{2}{x+1}=\dfrac{21}{10}$

    (c) $\dfrac{1}{x-1}-\dfrac{2}{x+1}=1$

8 For each of the following problems, form a quadratic equation, find its roots and write down the solution to the problem.

    (a) The height of a triangle is 10 cm more than the length of its base.
    The area of the triangle is 50 cm².
    What are the height and base length of the triangle?

    (b) The sum of the first n whole numbers, beginning at 1, is $\frac{1}{2}n(n+1)$. How many whole numbers are needed to add up to 465?

    (c) One side of a right-angled triangle is 5 cm less than the hypotenuse, which is twice as long as the other side.
    What are the lengths of the sides?

    (d) The **geometric mean** of two numbers is the square root of their product, e.g. geometric mean of 10 and 20 is
$$\sqrt{(10\times20)}=\sqrt{200}=14\cdot1 \text{ (to 3 significant figures)}$$

    The geometric mean of two numbers x and $(x+100)$ is $2x$.
    What are the two numbers?

REMINDERS AND REVISION — FUNCTIONS

Reminder facts

1 '$x>0$, $x \in R$' means 'real numbers more than zero'

'$x<1$, $x \in R$' means 'real numbers less than 1'

'$-2<x<3$, $x \in R$' means
'real numbers from -2 to 3, not including -2 and not including 3'

'$-5 \leqslant x \leqslant 5$, $x \in R$' means
'real numbers from -5 to 5, including -5 and including 5'

'$-1 \leqslant x < 4$, $x \in R$' means
'real numbers from -1 to 4, including -1 but not including 4'

2 $f(x)$ is said to be an **even** function of x if, when we change
the sign of a value of x, the value of $f(x)$ does not change,

e.g. $f(x)=2x^4$
$$f(3)=2 \times 3^4 = 2 \times 81 = 162$$
$$f(-3)=2 \times (-3)^4 = 2 \times 81 = 162$$

$f(x)$ is said to be an **odd** function of x if, when we change
the sign of a value of x, only the sign of the value of $f(x)$
changes,

e.g. $f(x)=4x^3$
$$f(2)=4 \times 2^3 = 4 \times 8 = 32$$
$$f(-2)=4 \times (-2)^3 = 4 \times -8 = -32$$

$f(x)$ is neither an even nor an odd function of x if, when we
change the sign of a value of x, the value of $f(x)$ changes,

e.g. $f(x)=(x+1)^2$
$$f(1)=(1+1)^2 = 2^2 = 4$$
$$f(-1)=(-1+1)^2 = 0^2 = 0$$

3 Another way of writing the function
$f(x)=2x+3$ $x \in R$
is as follows
$f : x \longrightarrow 2x+3$ $x \in R$,

You will find this notation used in some of the following
questions.

REVISION EXERCISE

1 For each of the following functions, make a
table, using integer values of x, and draw a
graph.

(a) $f(x)=x^2-x-2$ $-3 \leqslant x \leqslant 4$, $x \in R$
(b) $g(x)=6-x-x^2$ $-4 \leqslant x \leqslant 3$, $x \in R$
(c) $h(x)=2x^2+x-6$ $-3 \leqslant x \leqslant 3$, $x \in R$

2 Make a table, using integer values of x, and
draw a graph for each of the following
functions.

(a) $j : x \longrightarrow (x+2)(x-1)(x-3)$
    $-3 \leqslant x \leqslant 4$ $x \in R$

(b) $k : x \longrightarrow x^3+2x^2-x-2$
    $-3 \leqslant x \leqslant 3$ $x \in R$

177

3 In each of the following, caluclate f(1) and f(-1), i.e. the values of the function for $x=1$ and $x=-1$, and state whether the function is odd, even or neither even nor odd.

(a) $f(x)=x$ $x\in R$
(b) $f(x)=x-1$ $x\in R$
(c) $f(x)=\dfrac{1}{x}$ $x\neq 0,\ x\in R$
(d) $f(x)=x^2$ $x\in R$
(e) $f(x)=(x+3)^2$ $x\in R$
(f) $f(x)=\dfrac{1}{x^2}$ $x\neq 0,\ x\in R$
(g) $f(x)=2^x$ $x\in R$
(h) $f(x)=x^3$ $x\in R$
(i) $f(x)=(x-2)^3$ $x\in R$
(j) $f(x)=\dfrac{1}{x^3}$ $x\neq 0,\ x\in R$
(k) $f(x)=3^x$ $x\in R$

4 (a) Make a table of values for the function
$f(n)=(1\cdot 5)^n$
$n=4,\ 3,\ 2,\ 1,\ 0,\ -1,\ -2,\ -3,\ -4$

Note: We can calculate values of f(n) (to 1 decimal place) in two ways:
without a calculator
$(1\cdot 5)^4=\left(\dfrac{3}{2}\right)^4=\dfrac{3^4}{2^4}=\dfrac{81}{16}=5\cdot 1$

with a calculator
$(1\cdot 5)^4=5\cdot 1$ (easier and quicker!)

(b) Now draw a graph of the function
$f(x)=(1\cdot 5)^x$ $-4\leqslant x\leqslant 4$ $x\in R$

5 Break down each of the following functions into its parts.
Example $y=f(x)=4x+5$ $x\in R$

$$x\ \xrightarrow[\text{multiply by 4}]{\times 4}\ 4x\ \xrightarrow[\text{add 5}]{+5}\ 4x+5$$

(a) $y=g(x)=3x-2$ $x\in R$
(b) $y=h(x)=\frac{1}{4}x+3$ $x\in R$
(c) $y=j(x)=\dfrac{x}{6}-5$ $x\in R$
(d) $y=k(x)=\dfrac{x-3}{2}$ $x\in R$
(e) $y=m(x)=\frac{1}{3}(x+2)$ $x\in R$
(f) $y=n(x)=x^2+2$ $x\in R$
(g) $y=p(x)=3x^2$ $x\in R$
(h) $y=q(x)=3x^2+2$ $x\in R$

6 Complete (b) in the same way as (a)
(a) $y=f(x)=5x+4$ $x\in R$

$$x\ \xrightarrow[\text{multiply by 5}]{\times 5}\ 5x\ \xrightarrow[\text{add 4}]{+4}\ 5x+4$$

$$\dfrac{y-4}{5}\ \xleftarrow[\text{divide by 5}]{\div 5}\ y-4\ \xleftarrow[\text{subtract 4}]{-4}\ y$$

$$x=f^{-1}(y)=\dfrac{y-4}{5}$$

(b) $y=g(x)=2x-3$ $x\in R$

7 Find the inverse of each of the following functions.
(a) $y=f(x)=2x-1$ $x\in R$
(b) $y=g(x)=2(x-1)$ $x\in R$
(c) $y=h(x)=\frac{1}{3}x-4$ $x\in R$
(d) $y=j(x)=\dfrac{x}{5}+6$ $x\in R$
(e) $y=k(x)=\dfrac{x+2}{3}$ $x\in R$
(f) $y=m(x)=\frac{1}{2}(x-3)$ $x\in R$
(g) $y=n(x)=x^2-3$ $x\geqslant 0,\ x\in R$
(h) $y=p(x)=4x^2$ $x\geqslant 0,\ x\in R$
(i) $y=q(x)=4x^2-3$ $x\geqslant 0,\ x\in R$

REMINDERS AND REVISION — TRIG FUNCTIONS AND GRAPHS

Reminder facts

The sine, cosine or tangent of an angle can be got from a calculator,
e.g. $\sin 123° = 0\!\cdot\!839$ (to 3 significant figures)
$\quad \cos 231° = -0\!\cdot\!629$
$\quad \tan 312° = -1\!\cdot\!11$
$\quad \sin(-33°) = -0\!\cdot\!545$
$\quad \cos(415°) = 0\!\cdot\!574$

In each case, there is a connection between the angle in question and an acute angle (between 0° and 90°).

123°

acute angle = 57° (180° − 123°)
$\sin 123° = \sin 57° = 0\!\cdot\!839$
$\cos 123° = -\cos 57° = -0\!\cdot\!545$
$\tan 123° = -\tan 57° = -1\!\cdot\!54$

If A is between 90° and 180°
sin A is positive
cos A is negative
tan A is negative

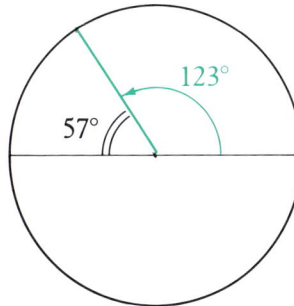

231°

acute angle = 51° (231° − 180°)
$\sin 231° = -\sin 51° = -0\!\cdot\!777$
$\cos 231° = -\cos 51° = -0\!\cdot\!629$
$\tan 231° = \tan 51° = 1\!\cdot\!23$

If A is between 180° and 270°
sin A is negative
cos A is negative
tan A is positive

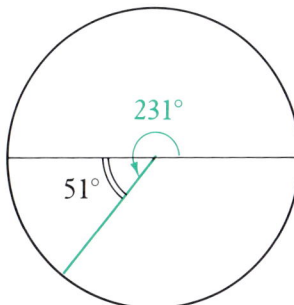

312°

acute angle = 48° (360° − 312°)
$\sin 312° = -\sin 48° = -0\!\cdot\!743$
$\cos 312° = \cos 48° = 0\!\cdot\!669$
$\tan 312° = -\tan 48° = -1\!\cdot\!11$

If A is between 270° and 360°
sin A is negative
cos A is positive
tan A is negative

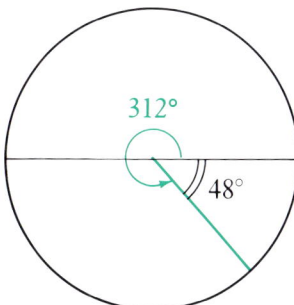

−33° is the same as **327°** $(360° − 33°)$

acute angle = 33°

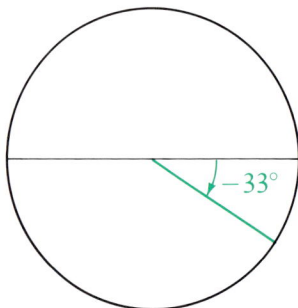

415° is the same as **55°** $(415° − 360°)$

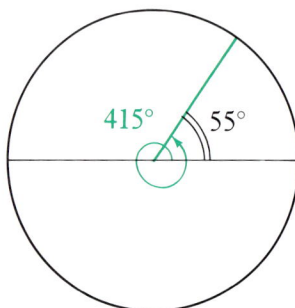

SUMMARY	sin A	cos A	tan A
A = 0°	0	1	0
0° < A < 90° (1st quadrant)	positive	positive	positive
A = 90°	1	0	not defined
90° < A < 180° (2nd quadrant)	positive	negative	negative
A = 180°	0	−1	0
180° < A < 270° (3rd quadrant)	negative	negative	positive
A = 270°	−1	0	not defined
270° < A < 360° (4th quadrant)	negative	positive	negative
A = 360°	0	1	0

Here are sketch graphs of some trigonometric functions.

$f(x) = \sin x \quad -720° \leqslant x \leqslant 720°$

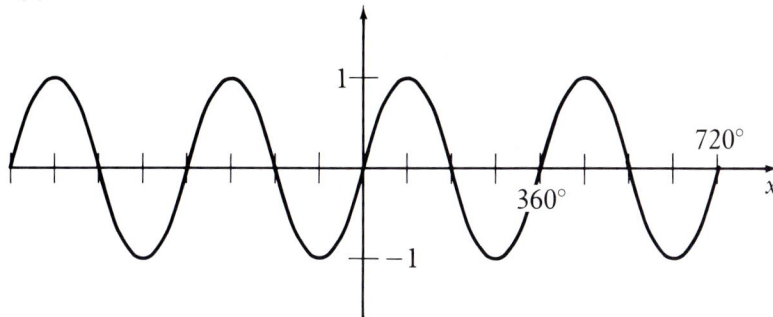

period $= 360°$, amplitude $= 1$

$f(x) = 2\sin x \quad -720° \leqslant x \leqslant 720°$

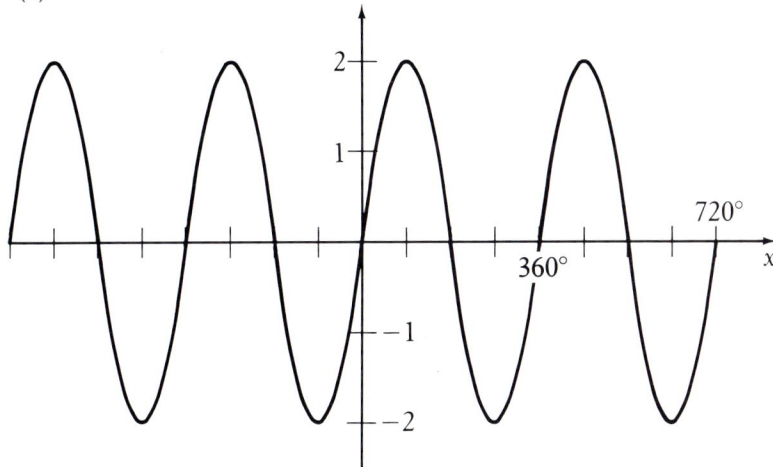

period $= 360°$, amplitude $= 2$

$f(x) = \sin 2x \quad -720° \leqslant x \leqslant 720°$

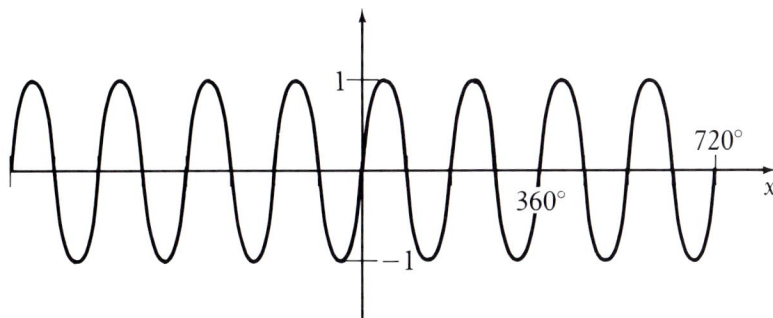

period $= 180°$, amplitude $= 1$

$f(x) = \cos x \quad -720° \leqslant x \leqslant 720°$

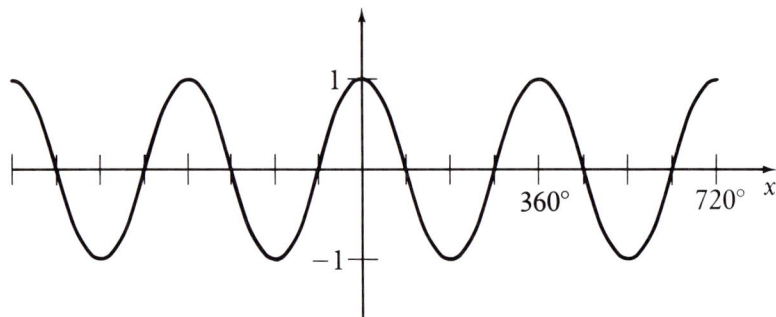

period = 360°, amplitude = 1

$f(x) = \cos\frac{1}{2}x \quad -720° \leqslant x \leqslant 720°$

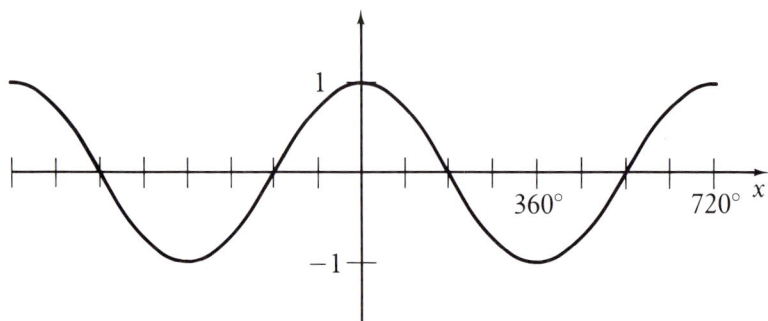

period = 720°, amplitude = 1

$f(x) = \cos(x + 180°) \quad -720° \leqslant x \leqslant 720°$

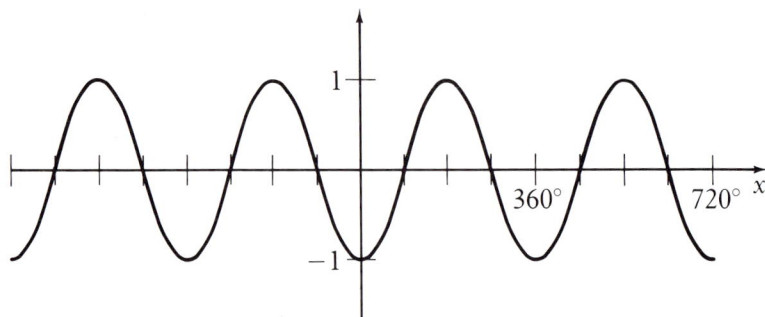

period = 360°, amplitude = 1

$f(x) = \sin x - \cos x \quad -720° \leqslant x \leqslant 720°$

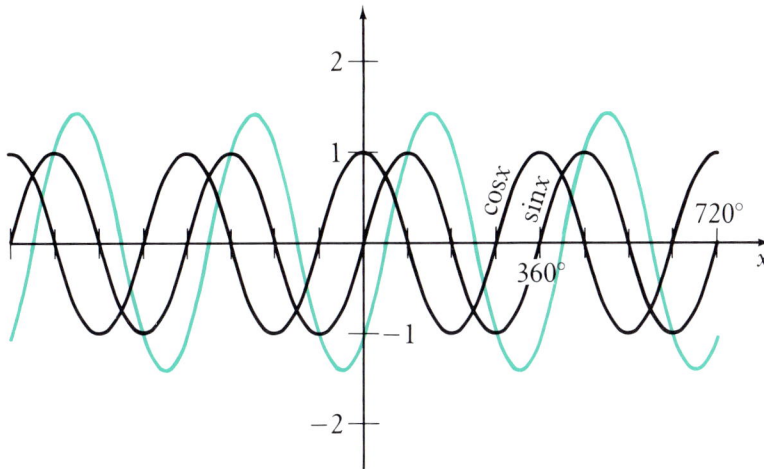

REVISION EXERCISE

1 (a) Copy and complete this table for an acute angle of 33°.
Give your answers to 3 significant figures.

	$\sin 33° = ?$
$\sin 147° = \sin ?°$	$\sin 147° = ?$
$\sin 213° = -\sin ?°$	$\sin 213° = ?$
$\sin 327° = -\sin ?°$	$\sin 327° = ?$

(b) Complete a similar table for an acute angle of 77°.

2 (a) Copy and complete this table for an acute angle of 44°.
Give your answers to 3 significant figures.

	$\cos 44° = ?$
$\cos 136° = -\cos ?°$	$\cos 136° = ?$
$\cos 224° = -\cos ?°$	$\cos 224° = ?$
$\cos 316° = \cos ?°$	$\cos 316° = ?$

(b) Complete a similar table for an acute angle of 66°.

In questions 3–7, graphs should be sketched for $-720° \leqslant x \leqslant 720°$

3 Sketch graphs of the following

(a) $1 - \cos x$
(b) $\sin x + 3$
(c) $\sin x + \cos x - 1$

4 Sketch graphs of the following

(a) $3\sin x$
What is the period and amplitude of this graph?

(b) $0.5\cos x$
What is the period and amplitude of this graph?

5 Sketch graphs of the following

(a) $\sin 3x$
What is the period and amplitude of this graph?

(b) $1.5\cos 2x$
What is the period and amplitude of this graph?

(c) $\cos \frac{1}{3}x$
What is the period and amplitude of this graph?

(d) $\frac{1}{4}\sin \frac{1}{2}x$
What is the period and amplitude of this graph?

6 Sketch graphs of the following

(a) $\sin(x - 90°)$
What is the period and amplitude of this graph?

(b) $\cos(x + 180°)$
What is the period and amplitude of this graph?

183

7 Sketch graphs of the following

    (a) $2\cos^2 x$

        What is the period and amplitude of this graph?

        Note: $2\cos^2 x$ means $(\cos x)^2 \times 2$.

    (b) $0{\cdot}5\sin^2 x$

        What is the period and amplitude of this graph?

8 (a) Copy and complete this table for an acute angle of 22°.

        Give your answers to 3 significant figures.

	$\tan 22° = ?$
$\tan 158° = -\tan ?°$	$\tan 158° = ?$
$\tan 202° = \tan ?°$	$\tan 202° = ?$
$\tan 338° = -\tan ?°$	$\tan 338° = ?$

    (b) Complete a similar table for an acute angle of 55°.

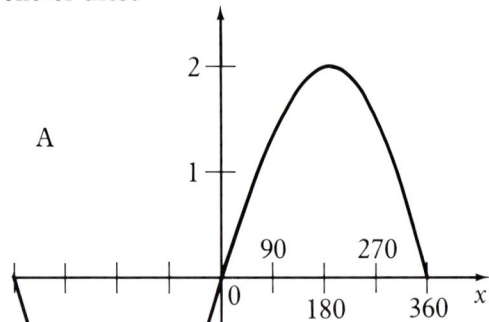

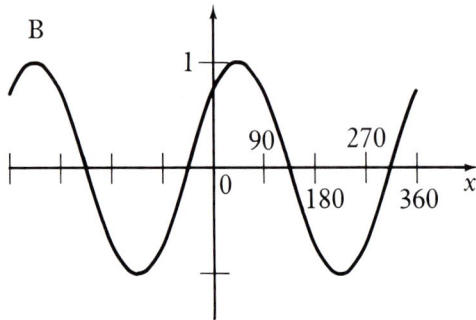

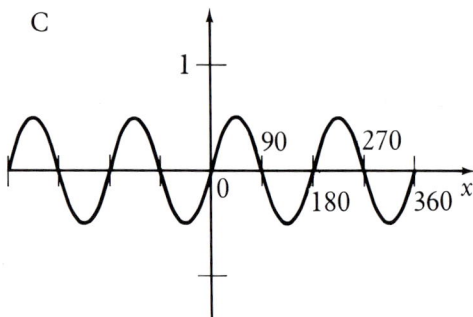

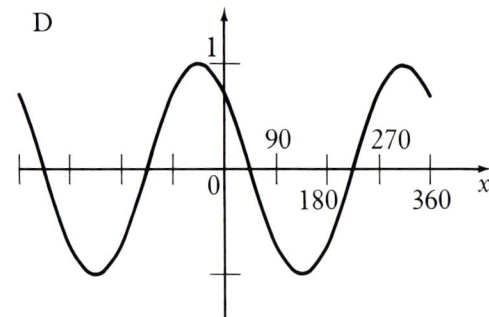

9 Do not use a calculator for this question.
Copy and complete the following table.

A	46°	52°	78°	102°	226°	308°
$\sin A$	0·719	0·788	0·978			
$\cos A$	0·695	0·616	0·208			
$\tan A$	1·036	1·280	4·705			

10 Each of the graphs in this question represents one of these functions.

$$f_1(x) = \tfrac{1}{2}\sin 2x \qquad -360° \leqslant x \leqslant 360°$$
$$f_2(x) = 2\sin \tfrac{1}{2}x \qquad -360° \leqslant x \leqslant 360°$$
$$f_3(x) = \cos(x+45) \qquad -360° \leqslant x \leqslant 360°$$
$$f_4(x) = \cos(x-45) \qquad -360° \leqslant x \leqslant 360°$$

Match each graph to its function

11 Each of the following graphs represents a trigonometric function. The function is written beside each graph in the form $a\sin(bx+c)$ or $a\cos(bx+c)$.

Write down the values of a, b, c as required for each.

(a) $a\cos bx$

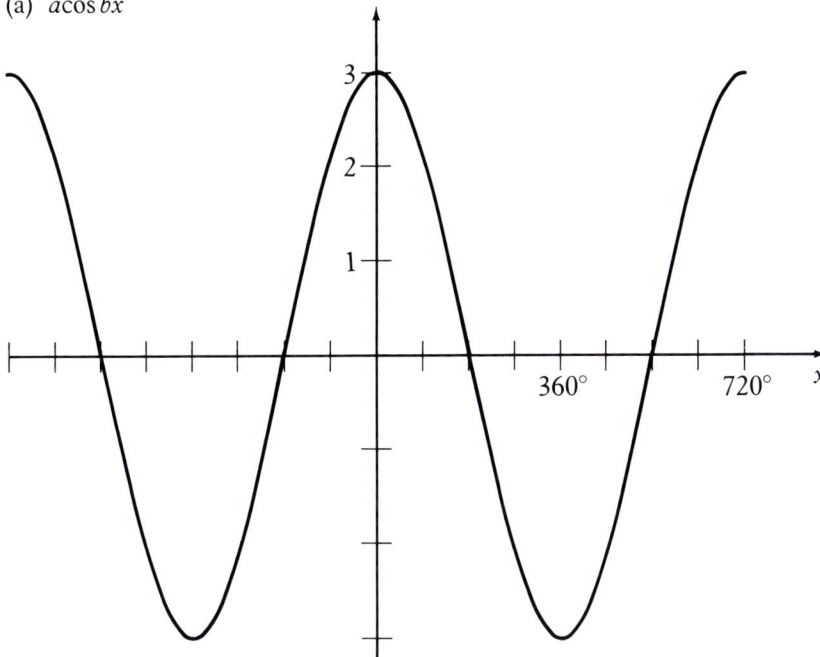

(b) $a\sin(x+c)$

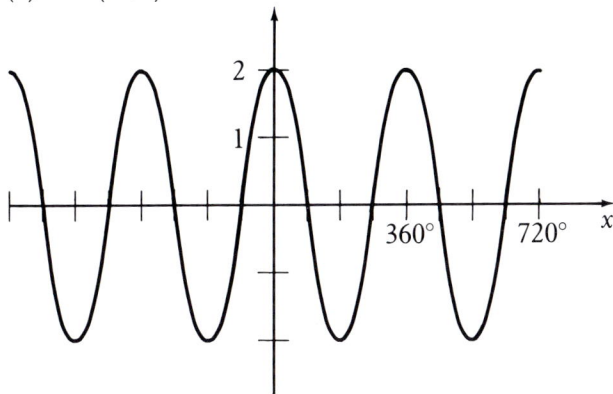

(c) $a\sin bx$

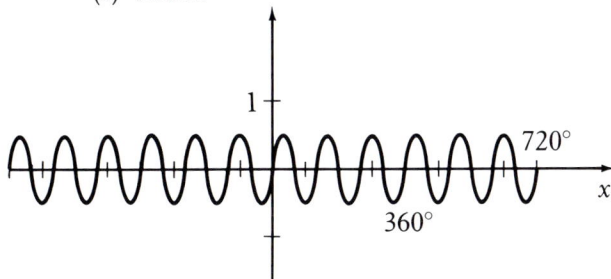

(d) $a\cos(x+c)$

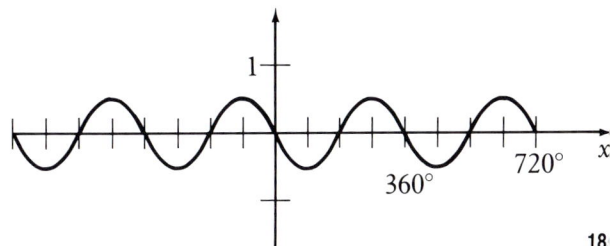

18

REMINDERS AND REVISION — PROPORTIONALITY

Reminder facts

DIRECT PROPORTIONALITY

p is directly proportional to q

$p \propto q$

$p = kq$ (where k is a constant)

or $\dfrac{p}{q} = k$

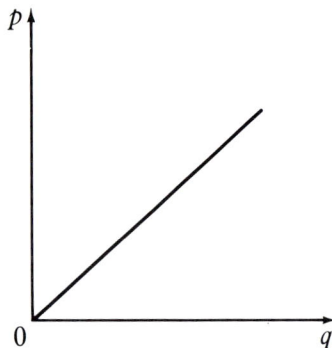

INVERSE PROPORTIONALITY

p is inversely proportional to q

$p \propto \dfrac{1}{q}$

$p = \dfrac{k}{q}$ (where k is a constant)

or $pq = k$

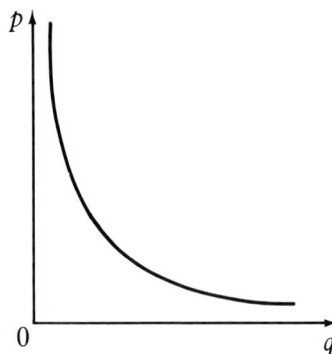

JOINT VARIATION

p is directly proportional to the square of q and inversely proportional to r.

$p \propto \dfrac{q^2}{r}$

$p = \dfrac{kq^2}{r}$ (where k is a constant)

REVISION EXERCISE

1 Indicate whether a table of data is an example of direct or inverse proportionality.

(a)

a	2	5	8
b	10	4	2·5

(b)

x	1	4	5
y	3	6	7

(c)

p	10	7·5	25
q	4	3	10

2 Select which graphs are examples of direct or inverse proportionality.

(a)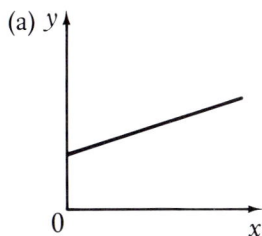

(b)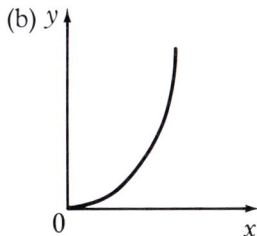

(c)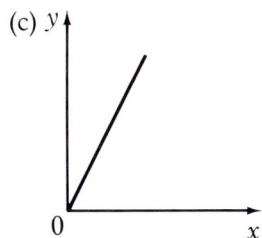

(d)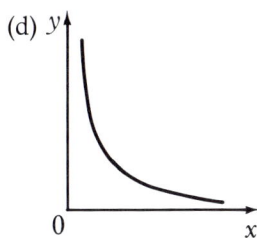

3 a is directly proportional to b.
$a = 18$ when $b = 36$.

(a) Find the formula for a in terms of b.
(b) Calculate a when $b = 27$.

4 p is inversely proportional to q.
$p = 10$ when $q = 3.5$.

(a) Find the formula for p in terms of q.
(b) Calculate q when $p = 7$.

5 l is directly proportional to m and inversely proportional to the square of n.
$l = 2$ when $m = 6$ and $n = 3$.

(a) Find the formula for l in terms of m and n.
(b) Calculate l when $m = 7$ and $n = 2$.
(c) Calculate n when $l = 6$ and $m = 50$.

6 The volume (V) of a given mass of gas is directly proportional to the temperature (T) and inversely proportional to the pressure (P).

Find the formula for V in terms of T and P.

7 The weight (W) of a cylinder is directly proportional to the length (l) and the square of the radius (r). $W = 3$ kg when $l = 60$ cm and $r = 4$ cm.

(a) Find the formula for W in terms of l and r.
(b) Calculate W when $l = 80$ cm and $r = 6$ cm.

8 The water resistance on a speed boat is directly proportional to the cube of the boat's speed.
(a) If the speed of the boat is doubled, what happens to the water resistance?
(b) If the water resistance has doubled, what has happened to the speed of the boat?

REMINDERS AND REVISION — TRIGONOMETRY

Reminder facts

USING TRIG RATIOS IN TRIANGLES WITHOUT A RIGHT ANGLE

In triangles without a right angle it is often possible to use one of the trig ratios by forming two right-angled triangles, e.g.

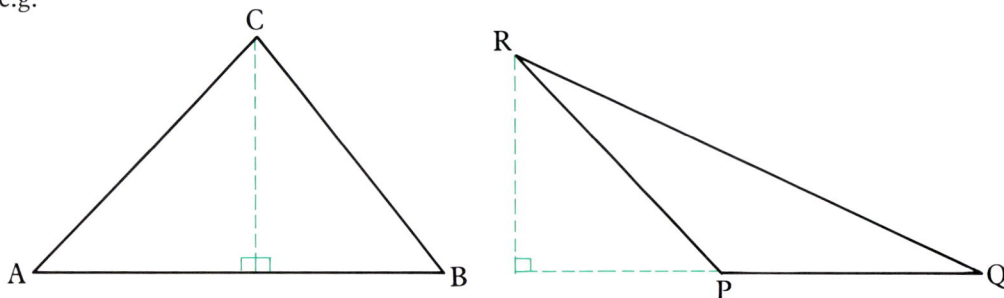

SINE RULE

$$\frac{a}{\sin A} = \frac{b}{\sin B} = \frac{c}{\sin C}$$

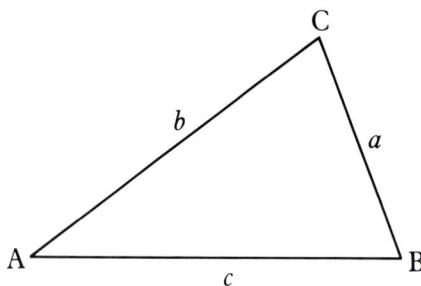

COSINE RULE

$$a^2 = b^2 + c^2 - 2bc\cos A$$

OR

$$\cos A = \frac{b^2 + c^2 - a^2}{2bc}$$

AREA OF A TRIANGLE

Area of triangle $ABC = \frac{1}{2}bc\sin A$

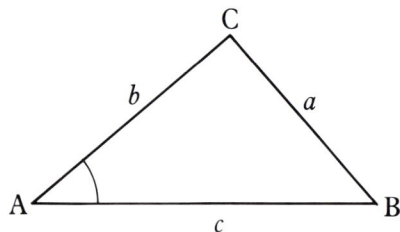

RELATIONSHIPS

(a) $\tan A = \dfrac{\sin A}{\cos A}$

(b) $\sin^2 A + \cos^2 A = 1$

REVISION EXERCISE

All answers should be given to two decimal places.

1 Calculate the length of PR.

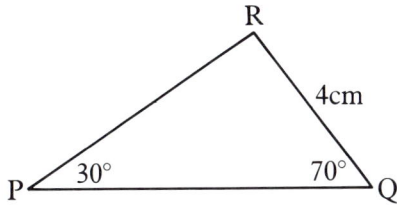

2 Calculate the size of the angle at L.

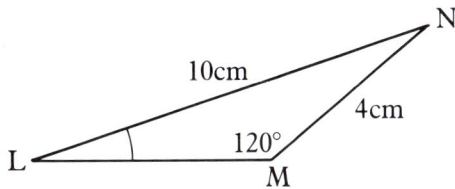

3 Calculate the length of AB.

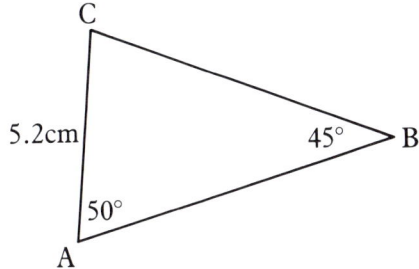

4 Calculate the size of angle C.

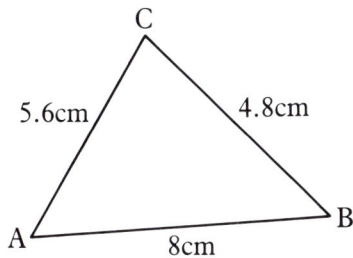

5 Calculate the length of QR.

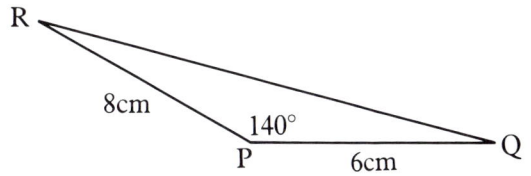

6 Calculate the smallest angle.

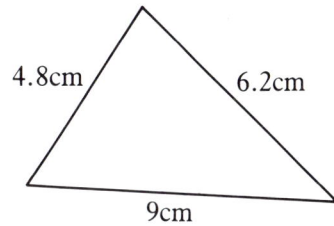

7 Calculate the area of triangle LMN.

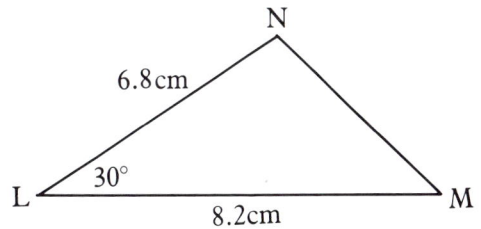

8 Calculate the area of triangle ABC.

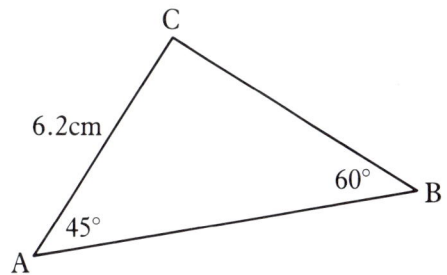

9 A sailor knows that the Church steeple is due north of the cairn and that the distance between them is 500 m.

The bearing of the Church steeple from the yacht is 310° and the bearing of the cairn from the yacht is 260°

Calculate the distance between the yacht and the Church steeple.

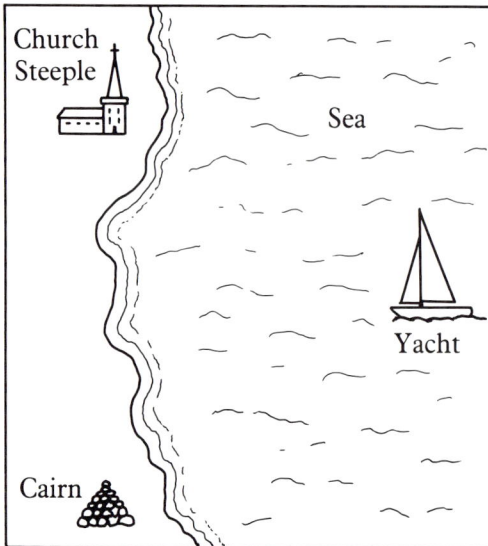

10 A, B, C, D, P, Q, R, S is a cube of side 5 cm.

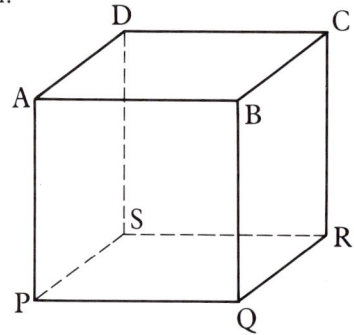

(a) Calculate the length of AC.
(b) Calculate the size of angle PRQ.

11 A, B, C, D, E, F, G is a cuboid of length 10 cm, height 7 cm and width 6 cm.

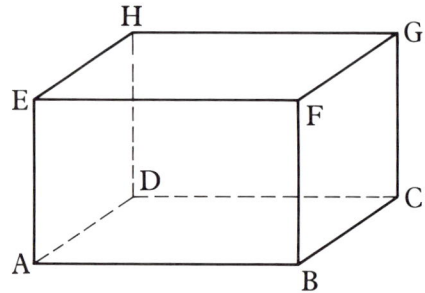

(a) Calculate the length of EC.
(b) Calculate the size of angle FCG.
(c) Calculate the size of angle ECA.

12 Prove that:

(a) $(\sin A + \cos A)^2 = 1 + 2\sin A \cos A$

(b) $\dfrac{\cos A}{\sin A} \times \tan A = 1$

13 Solve the equations to the nearest degree, for $0 \leqslant a \leqslant 360$

(a) $\cos a^0 = 0{\cdot}5$
(b) $2\tan a^0 = 0{\cdot}4$
(c) $5\sin a^0 + 2 = \sin a^0$
(d) $\tan^2 a^0 = 64$

REMINDERS AND REVISION — TRANSFORMATIONS AND MATRICES

Reminder facts

TRANSFORMATION EQUATIONS

You will find below a summary of the transformation equations that we obtained in the unit.

Before you use the summary, cover up the equations column and try to obtain as many of the transformation equations as you can by plotting a few points on a sheet of squared paper and 'doing' each of the transformations. You should be able to work out the equations for at least some of the transformations.

TRANSFORMATION	**EQUATIONS**
reflection in the line $y=x$	$x'=y$ $y'=x$
reflection in the line $y=-x$	$x'=-y$ $y'=-x$
reflection in the x-axis	$x'=x$ $y'=-y$
reflection in the y-axis	$x'=-x$ $y'=y$
reflection in the line $x=k$	$x'=2k-x$ $y'=y$
reflection in the line $y=j$	$x'=x$ $y'=2j-y$
rotation about O through $180°$ (half-turn)	$x'=-x$ $y'=-y$
rotation about O through $90°$ (quarter-turn anticlockwise)	$x'=-y$ $y'=x$
rotation about O through $-90°$ (quarter-turn clockwise)	$x'=y$ $y'=-x$
translation $\begin{pmatrix} a \\ b \end{pmatrix}$	$x'=x+a$ $y'=y+b$
dilatation [O,k]	$x'=kx$ $y'=ky$

EXAMPLES

(a) $(2,3)$ reflected in the line $y=x$: $\quad x'=y=3$
$$y'=x=2$$

$(2,3) \longrightarrow (3,2)$

(b) $(-3,4)$ reflected in the x-axis: $\quad x'=x=-3$
$$y'=-y=-4$$

$(-3,4) \longrightarrow (-3,-4)$

(c) $(1,-2)$ reflected in the line $y=3$: $\quad x'=x=1$
$$y'=2\times3-y$$
$$=6-(-2)=8$$

$(1,-2) \longrightarrow (1,8)$

(d) $(3,1)$ rotated about O through $180°$: $\quad x'=-x=-3$
$$y'=-y=-1$$

$(3,1) \longrightarrow (-3,-1)$

TRANSFORMATION MATRICES

We obtained matrices for some of the transformation equations above.

TRANSFORMATION	EQUATIONS	MATRIX
reflection in the line $y=x$	$x'=y$ $y'=x$	$\begin{pmatrix} 0 & 1 \\ 1 & 0 \end{pmatrix}$
reflection in the line $y=-x$	$x'=-y$ $y'=-x$	$\begin{pmatrix} 0 & -1 \\ -1 & 0 \end{pmatrix}$
reflection in the x-axis	$x'=x$ $y'=-y$	$\begin{pmatrix} 1 & 0 \\ 0 & -1 \end{pmatrix}$
reflection in the y-axis	$x'=-x$ $y'=y$	$\begin{pmatrix} -1 & 0 \\ 0 & 1 \end{pmatrix}$
rotation about O through $180°$ (half-turn)	$x'=-x$ $y'=-y$	$\begin{pmatrix} -1 & 0 \\ 0 & -1 \end{pmatrix}$
rotation about O through $90°$ (quarter-turn anticlockwise)	$x'=-y$ $y'=x$	$\begin{pmatrix} 0 & -1 \\ 1 & 0 \end{pmatrix}$
rotation about O through $-90°$ (quarter-turn clockwise)	$x'=y$ $y'=-x$	$\begin{pmatrix} 0 & 1 \\ -1 & 0 \end{pmatrix}$

EXAMPLES

(a) $(2,3)$ reflected in the line $y=-x$: $\quad \begin{pmatrix} x' \\ y' \end{pmatrix} = \begin{pmatrix} 0 & -1 \\ -1 & 0 \end{pmatrix}\begin{pmatrix} 2 \\ 3 \end{pmatrix} = \begin{pmatrix} -3 \\ -2 \end{pmatrix}$

$(2,3) \longrightarrow (-3,-2)$

(b) $(2,3)$ reflected in the y-axis: $\quad \begin{pmatrix} x' \\ y' \end{pmatrix} = \begin{pmatrix} -1 & 0 \\ 0 & 1 \end{pmatrix}\begin{pmatrix} 2 \\ 3 \end{pmatrix} = \begin{pmatrix} -2 \\ 3 \end{pmatrix}$

$(2,3) \longrightarrow (-2,3)$

(c) $(2,3)$ rotated about O through $90°$: $\quad \begin{pmatrix} x' \\ y' \end{pmatrix} = \begin{pmatrix} 0 & -1 \\ 1 & 0 \end{pmatrix}\begin{pmatrix} 2 \\ 3 \end{pmatrix} = \begin{pmatrix} -3 \\ 2 \end{pmatrix}$

$(2,3) \longrightarrow (-3,2)$

Double transformations

M_1 is the matrix for a transformation, T_1,
M_2 is the matrix for a transformation, T_2 and
(x, y) is the point being transformed.

We can obtain the image point (x'', y'') after the double
transformation T_1 followed by T_2 as follows

$$\begin{pmatrix} x'' \\ y'' \end{pmatrix} = M_2 M_1 \begin{pmatrix} x \\ y \end{pmatrix}$$

The matrix for the double transformation T_1 followed by T_2
is the product of the two matrices $M_2 M_1$.

Notice the order of the matrices in the product $M_2 M_1$.

EXAMPLE

A quarter-turn anticlockwise followed by a half-turn

$$M_1 = \begin{pmatrix} 0 & -1 \\ 1 & 0 \end{pmatrix} \qquad M_2 = \begin{pmatrix} -1 & 0 \\ 0 & -1 \end{pmatrix}$$

The product of the two matrices is

$$\begin{pmatrix} -1 & 0 \\ 0 & -1 \end{pmatrix} \begin{pmatrix} 0 & -1 \\ 1 & 0 \end{pmatrix} = \begin{pmatrix} 0 & 1 \\ -1 & 0 \end{pmatrix} \qquad \text{the matrix for } \textit{a quarter-turn clockwise}$$

A quarter-turn anticlockwise followed by a half-turn
has the same effect as
a quarter-turn clockwise

REVISION EXERCISE

Transformation equations

1 Consider a triangle with vertices A(1,4),
B(-3,2), C(0,-5).

Write down the coordinates of the vertices
of the image of triangle ABC after each of
the following transformations.

(a) reflection in the line $y = x$
(b) reflection in the line $y = -x$
(c) reflection in the x-axis
(d) reflection in the y-axis
(e) reflection in the line $x = 2$
(f) reflection in the line $y = -3$
(g) rotation about O through 180°
    (half-turn)
(h) rotation about O through 90°
    (quarter-turn anticlockwise)
(i) rotation about O through $-90°$
    (quarter-turn clockwise)
(j) translation $\begin{pmatrix} 1 \\ -2 \end{pmatrix}$

(k) dilatation [O,2]

Transformation matrices

2 In each of the following,
  (i) work out the double transformation
      matrix,
 (ii) state which single transformation has the
      same effect.

(a) a half-turn followed by a quarter-turn
    clockwise
(b) a quarter-turn anticlockwise followed by
    a half-turn
(c) reflection in $y = -x$ followed by reflection
    in $y = x$
(d) reflection in the y-axis followed by
    reflection in the x-axis
(e) reflection in the x-axis followed by
    reflection in the line $y = x$
(f) reflection in the y-axis followed by
    reflection in the line $y = x$
(g) reflection in the x-axis followed by
    reflection in the line $y = -x$
(h) reflection in the y-axis followed by
    reflection in the line $y = -x$.

REMINDERS AND REVISION — STATISTICS

Reminder facts

The **mean** is what is commonly called 'the average', and is calculated by adding all the values and dividing by the number of values.

For example: the mean of 3, 0, 2, 8, 9, 5 is
$$\frac{3+0+2+8+9+5}{6}=\frac{27}{6}=4\cdot5$$

The **median** is the central value when a set of values are arranged in order.

For example: the order of 4, 2, 1, 15, 3, 2, 4 is given by
1, 2, 2, 3, 4, 4, 15.
The median is 3.

The **mode** of a set of values is the most frequently occurring value.

For example: the mode of 4, 1, 4, 3, 10, 1, 2, 4, 3 is 4.

Cumulative frequency is the 'running total' of frequency.

The frequency table below gives the marks for a class of 30 students in an examination.

Marks	Frequency	Cumulative frequency	Less than
0– 5	1	1	5
6–10	3	$1+3=\ 4$	10
11–15	6	$4+6=10$	15
16–20	8	$10+8=18$	20
21–25	7	$18+7=25$	25
26–20	5	$25+5=30$	30

Cumulative frequency is used to indicate the number of items that are less than a stated value.

The **range** of a set of values is the difference between the smallest and the largest values.

Range = largest value − smallest value.

Quartiles — best described by giving an example.

Example: the weights (kg) of 20 students are given below in a stem-leaf table.

Weight (kg)

```
3 | 1 2 2 5 7
4 | 1 1 4 7 8 8 9
5 | 0 3 4 8 9
6 | 1 1 2
```

Key   3 | 1 = 31 kg

The median is between the 10th and 11th weight,
i.e. between 48 kg and 48 kg.
Median is 48 kg.

The lower quartile comes between the 5th and 6th weight,
i.e. between 37 kg and 41 kg.
The lower quartile is 39 kg.

The upper quartile comes between the 15th and 16th
weight, i.e. between 54 kg and 58 kg.
The upper quartile is 56 kg.

The **interquartile range** is the difference between the
upper and lower quartile.
Interquartile range = upper quartile − lower quartile.

REVISION EXERCISE

1 Find the mean, median and the mode of the following sets of values.

(a) 7, 2, 5, 3, 6, 3, 9
(b) 11, 4, 2, 2, 4, 5, 6, 2
(c) 2·8 m, 2·4 m, 3·1 m, 2·6 m, 2·6 m

2 The mean of 4, 7, 8 and x is 6. What is the value of x?

3 The typing speeds (in words per minute) for 25 students are given below in a stem-leaf table.

Words per minute

```
3 | 8 8 9 9
4 | 0 1 1 8 9 9
5 | 0 0 2 2 2 3 3 4 6 9
6 | 0 0 1 1 3
```

Find the mean, median and mode (to 1 decimal place, where necessary).

4 The heights (m) of 20 students were recorded as follows:

1·6 1·7 1·5 1·8 1·7
1·9 1·4 1·6 1·8 1·7
1·3 1·4 1·6 1·7 1·6
1·4 1·5 1·6 1·4 1·6

(a) Find the mean, median and mode (to 2 decimal places, where necessary).
(b) Calculate the upper and lower quartiles (to 2 decimal places, where necessary).

5 The scores of 30 golfers in the first round of a competition were as follows:

Scores	Frequency
60–64	1
65–69	7
70–74	10
75–79	6
80–84	4
85–90	2

(a) Calculate the mean score from the data given (to the nearest whole number).

(b) 60% of the golfers (with the lowest scores) go through to the next round. By considering cumulative frequencies, work out the maximum score to qualify for round 2.

6 A care hire rental company recorded the daily mileage of 50 of its cars. These were as follows:

```
 92   97  103  125  149
 97  109   87  115  132
118  106   89  113  102
109   97  113   87  130
146   82  113   97  106

105   83  150  141  132
121  111  116   97  141
131   83  117  148  112
141  117  112  103  115
115  109  142  127  119
```

By constructing a stem-leaf table or a cumulative frequency diagram,
(a) find the median,
(b) find the upper and lower quartiles,
(c) calculate the interquartile range.

ANSWERS

Unit 1. Factors

1 (a) $4(a+2b)$ (b) $3(2c-d)$ (c) $4(x+2)$ (d) $4(a+1)$ (e) $2(3x-y)$ (f) $3(b-2)$

2 (a) $2b(a+c)$ (b) $3p(q-2r)$ (c) $3x^2(3y-z)$ (d) $5qr(p-2)$ (e) $4xy(y-2x)$ (f) $ab(7a-3b)$

   (g) $4lm(2lm+1)$ (h) $7xy(3x-1)$ (i) $xy^2z(x-z)$ (j) $lmn(lmn-1)$

3 (a) 2 (b) $-\frac{4}{7}$ (c) $-\frac{1}{5}$ (d) $-\frac{5}{4}$ (e) $-\frac{1}{3}$ (f) $\frac{7}{5}$

4 (a) $10x^2-4xy$ (b) $15xy-6y^2$ (c) $2x^2+3xy$ (d) $-4xy-6y^2$

5 (a) $6x^2+13xy+6y^2$ (b) $3p^2+5pq-2q^2$ (c) $4l^2-10lm-6m^2$ (d) $2x^2-5xy+2y^2$

   (e) $12a^2-7ab-12b^2$ (f) $5c^2-21cd+4d^2$

6 (a) a^2+3a+2 (b) $b^2+8b+15$ (c) $p^2-10p+24$ (d) y^2-4y+3 (e) r^2+7r+6 (f) c^2-6c+9

7 (a) $2x^2+5xy+3y^2$ (b) $2a^2-3ab+b^2$ (c) $2p^2+3pq-2q^2$ (d) $9l^2-4m^2$ (e) $15x^2-19x+6$

   (f) $1-2a-3a^2$ (g) $20-37x+15x^2$ (h) $ab+a+b+1$ (i) $6-3x-2y+xy$

8 (a) $(x+4)(x+1)$ (b) $(a+2)(a+1)$ (c) $(b+4)(b+2)$ (d) $(p+7)(p+3)$ (e) $(q+3)(q+1)$

   (f) $(x+1)(x+1)$

9 (a) $(x-3)(x-2)$ (b) $(p-4)(p-2)$ (c) $(p-3)(p-3)$ (d) $(x-5)(x-1)$ (e) $(q-7)(q-3)$ (g) $(q-8)(q-2)$

   (g) $(m+4)(m+1)$ (h) $(y+5)(y+2)$ (i) $(a+1)(a+3)$

10 (a) $(x+3)(x-2)$ (b) $(x+3y)(x-2y)$ (c) $(a-3)(a+2)$ (d) $(a-3b)(a+2b)$ (e) $(l+3m)(l-m)$

   (f) $(p+8q)(p-2q)$ (g) $(c-3d)(c-4d)$ (h) $(r+5)(r-1)$ (i) $(p+2q)(p+3q)$

11 (a) $(2+3x)(1-2x)$ (b) $(2+3a)(3-a)$ (c) $(3a+2)(a-3)$ (d) $(5x-3)(x+4)$ (e) $(3a+2)(2a-3)$

   (f) $(4p+7)(2p-1)$ (g) $(3q-13)(2q+1)$ (h) $(3x-2y)(2x+3y)$ (i) $(4a-3b)(2a+b)$ (j) $(3-2y)(2-3y)$

   (k) $(2+4r)(1+5r)$ (l) $(3c-4d)(2c+5d)$

12 (a) $(a-b)(a+b)$ (b) $(p-2q)(p+2q)$ (c) $(3c-d)(3c+d)$ (d) $(4x-5y)(4x+5y)$

   (e) $(2p-q)(2p+q)$ (f) $(1-4n)(1+4n)$ (g) $(a-b)(a+b)(a^2+b^2)$ (h) $10(2m-1)(2m+1)$

   (i) $-2(3b-7)(3b+7)$

13 (a) $(x+1)(x+1)$ (b) $(p-2q)(p+2q)$ (c) $(a-3)(a-4)$ (d) $(p+3)(2p-1)$ (e) $(3a-2)(4a-1)$

   (f) $2(2a+3b)(2a+b)$ (g) $2(x+3y)(2x-y)$ (h) $3(p-q)(2p-3q)$ (i) $3(m-3)(m+3)$ (j) $3(x-2)(2x+1)$

   (k) $7(1-2m)(1+2m)$ (l) $3(3p+2q)^2$

14 (a) $4p+q$ (b) $\frac{1}{4}$ (c) $x+y$ (d) $\dfrac{a-3}{3}$ (e) $\dfrac{a+2}{a+3}$ (f) $\dfrac{p+4q}{p-4q}$

REVISION—FACTORS

1 (a) x^2+5x+6 (b) $10p^2-7pq+q^2$ (c) $13mn-6m^2-6n^2$ (d) $8r^2+14rs-15s^2$

2 (a) $4mn(2m-3n)$ (b) $(x+3)(x+7)$ (c) $2(p-q)(p+3q)$ (d) $(6y-7z)(6y+7z)$

   (e) $def(d^2e+e^2f-df^2)$ (f) $(1-2r)(2-r)$ (g) $(3a+2b)(4a-3b)$ (h) $3(3-2c)(3+2c)$

   (i) $2(x^2+x+2)$ (j) $5(2s-3t)(3s+7t)$ (k) $(4c-5d)(3c-4d)$ (l) $(1-10z)(1+10z)$

3 (a) $\dfrac{2}{x+1}$ (b) $\dfrac{x+2y}{x-2y}$ (c) $\dfrac{2(2p+3)}{3p+2}$ (d) $\dfrac{1}{1+a}$

4 Area of square $-$ area of rectangle $=L^2-L(L-2)=L^2-(L^2-2L)=2L$

5 Area of square $-$ area of rectangle $=L^2-(L-2)(L+2)=L^2-(L^2-4)=4$

6 Area of square $-$ area of rectangle $=L^2-(L+2)(L-3)=L^2-(L^2-L-6)=L+6$

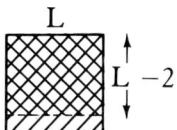
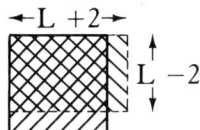
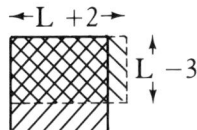

Unit 2.   Equation of a straight line and inequalities

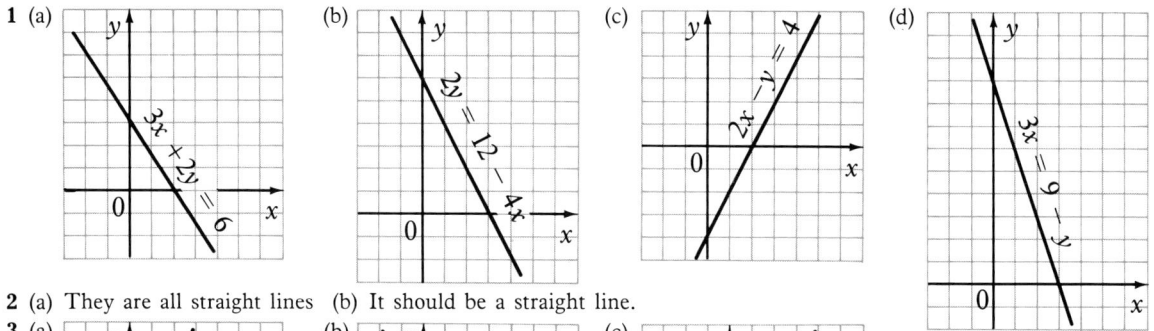

1 (a)

(b)

(c)

(d)

2 (a) They are all straight lines   (b) It should be a straight line.

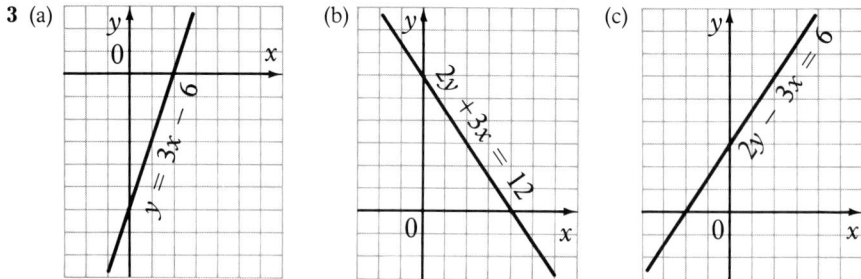

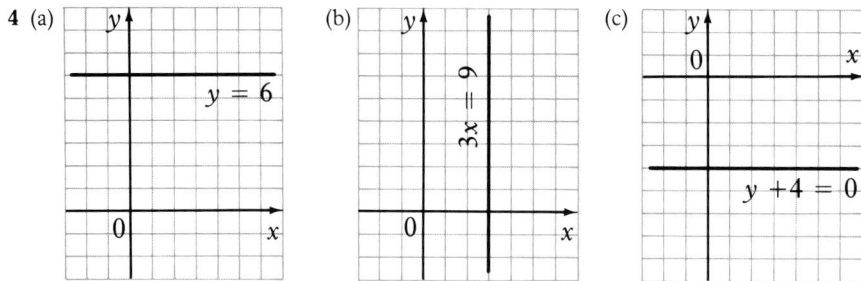

3 (a)

(b)

(c)

4 (a)

(b)

(c)

5 Straight lines parallel to either the x or y-axis

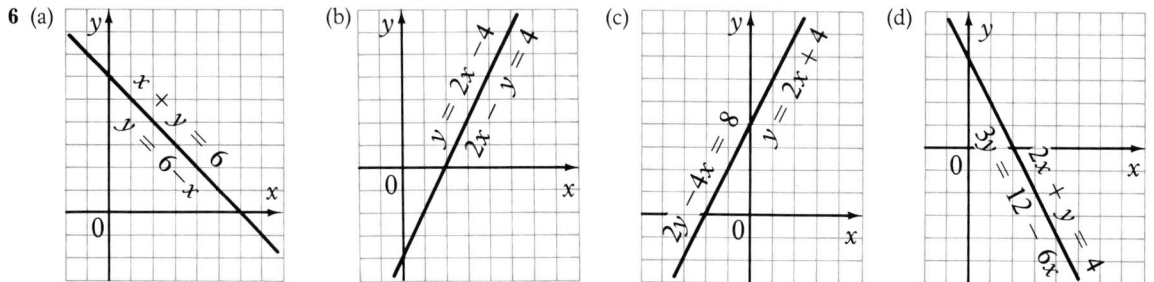

6 (a)

(b)

(c)

(d)

7 The two equations in each pair are just different forms of the same equation.

8 $x+y=4$ and $y=4-x$ $2x-y=7$ and $y=2x-7$ $2y=14-4x$ and $y+2x=7$ $2y=2x+4$ and $y-x=2$

9 (a) $y=-2x+6$ (b) $y=3x+6$ (c) $y=-2x+4$ (d) $y=x-7$ (e) $y=-\frac{3}{2}x+4$ (f) $y=2x-3$

10 (a) gradient $=1$, intercept $=4$ (b) gradient $=2$, intercept $=0$ (c) gradient $=\frac{1}{2}$, intercept $=-2$

11 (a) gradient $=-2$, intercept $=4$ (b) gradient $=3$, intercept $=-5$ (c) gradient $=-3$, intercept $=4$
   (d) gradient $=-2$, intercept $=0$

12 gradient $=a$, intercept $=b$

13 (a) gradient $=3$, intercept $=-6$ (b) gradient $=-\frac{1}{2}$, intercept $=4$ (c) gradient $=-3$, intercept $=6$
   (d) gradient $=\frac{1}{2}$, intercept $=3$

14 (a) gradient $=-1$, intercept $=-4$ (b) gradient $=\frac{1}{2}$, intercept $=-3\frac{1}{2}$
   (c) gradient $=-1\frac{1}{2}$, intercept $=-2$ (d) gradient $=3$, intercept $=7$

15 (a) $y = 2x + 4$ (b) $y = -3x - 4$ (c) $y = \frac{1}{2}x$ (or $2y = x$)

16 (a) points such as $(1,6), (2,8)$ (b) points such as $(1,-7), (2,-10)$ (c) points such as $(2,1), (4,2)$

17 (a) $y = -x + 4$ (b) $y = -2x + 4$ (c) $y = -\frac{1}{2}x + 1$ (d) $y = 3x - 7$

18 For example, (a) $x + y = 4$ (b) $2x + y = 4$ (c) $x + 2y = 2$ (d) $3x - y = 7$

19 (a) (b) They are parallel

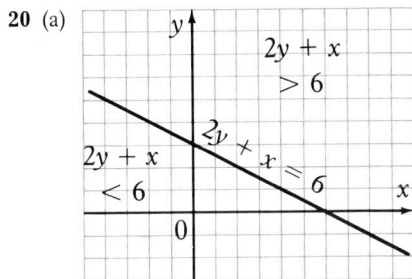

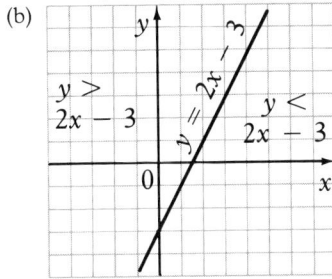

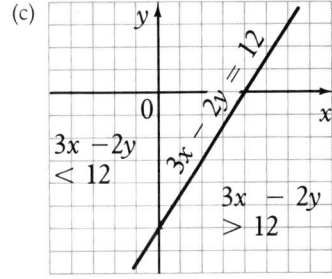

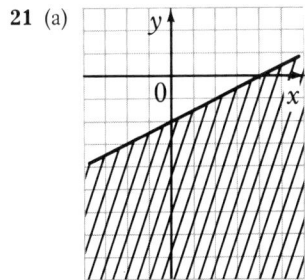

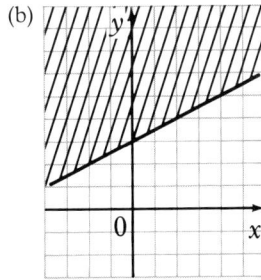

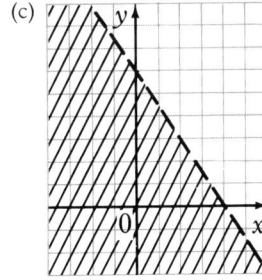

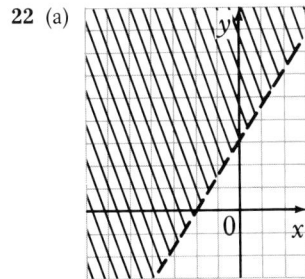

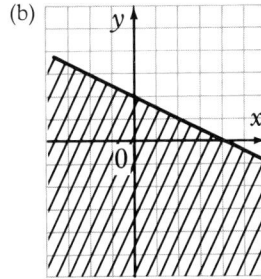

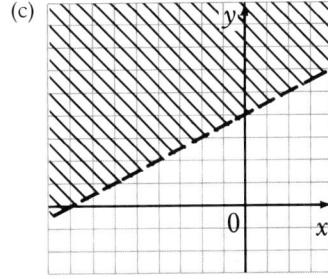

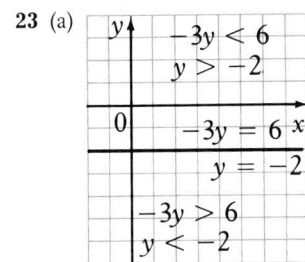

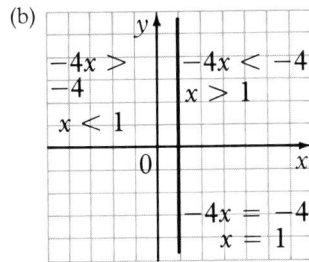

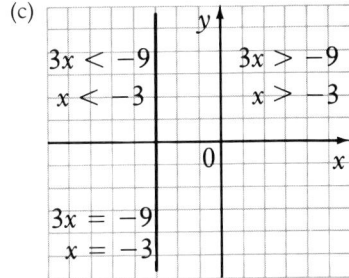

$y = 2x + 4$, $y = 2x - 2$

$y = \frac{1}{2}x + 5$, $y = \frac{1}{2}x - 3$

20 (a)

$2y + x > 6$, $2y + x = 6$, $2y + x < 6$

(b)

$y > 2x - 3$, $y = 2x - 3$, $y < 2x - 3$

(c)

$3x - 2y = 12$, $3x - 2y < 12$, $3x - 2y > 12$

21 (a) (b) (c)

22 (a) (b) (c)

23 (a)

$-3y < 6$, $y > -2$
$-3y = 6$
$y = -2$
$-3y > 6$, $y < -2$

(b)

$-4x > -4$, $x < 1$
$-4x < -4$, $x > 1$
$-4x = -4$, $x = 1$

(c)

$3x < -9$, $x < -3$
$3x > -9$, $x > -3$
$3x = -9$, $x = -3$

24 (a) $x<2$ (b) $x<5$ (c) $x\leqslant 3$ (d) $x<6$ (e) $x<4$ (f) $x>2$ (g) $x>-15$ (h) $x\leqslant -2$

25 (a) $\{x : x<-1, x\in Z\}$ (b) $\{x : x<-1\cdot 6, x\in R\}$ (c) $\{x : x<4, x\in Z\}$
   (d) $\{x : x>-1\cdot 625, x\in R\}$ (e) $\{x : x<-1, x\in Z\}$ (f) $\{x : x<2, x\in Z\}$

Revision—Straight line and inequalities

1 (a) gradient $=2$, intercept $=-3$ (b) gradient $=-3$, intercept $=4$ (c) gradient $=-2$, intercept $=0$
   (d) gradient $=-\frac{1}{2}$, intercept $=3$ (e) gradient $=3$, intercept $=5$ (f) gradient $=-\frac{2}{3}$, intercept $=-4$

2 (a) $y=3x+4$ (b) $y=-2x$ (c) $y=-\frac{1}{2}x-2$ **3** $y=2x-2$

4 (a) $y=x$ (b) $y=x-2$ (c) $2y=x-2$ (d) $y=1-2x$ **5** (a) $x\leqslant 4$ (b) $y>-x$ (c) $x+2y\leqslant 6$

6 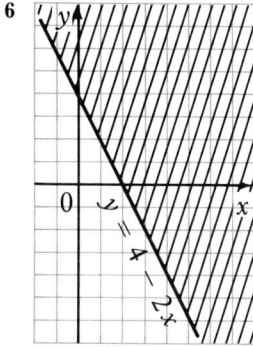 **7**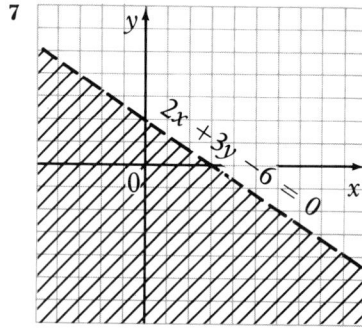

8 (a) $x>2$ (b) $x\leqslant 3$ (c) $x\geqslant 3$ (d) $x\geqslant -6$ (e) $x\geqslant -3$

Unit 3. Simultaneous equations & linear programming

1 (a) $x=3$, $y=1$ (b) $x=-1$, $y=-1$

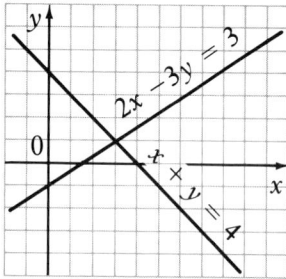 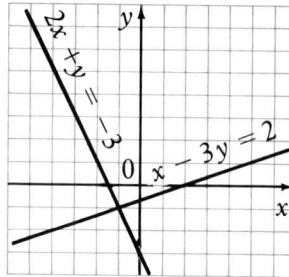

2 (a) $x=2\cdot 5$, $y=3\cdot 5$ (correct to 1 decimal place) (b) $x=1$, $y=-1\cdot 5$ (correct to 1 decimal place)

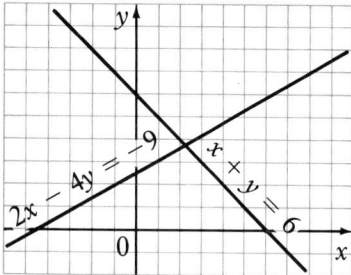 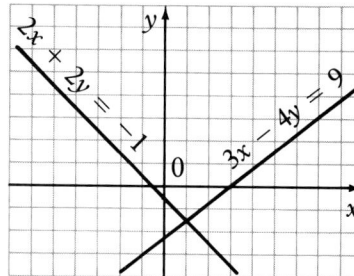

3 The lines are parallel.

4 (a) $x=6$, $y=1$ (b) $x=1$, $y=2$ (c) $a=3$, $b=-3$ (d) $a=5$, $b=2$ (e) $x=-\frac{1}{3}$, $y=\frac{4}{3}$ (f) $a=4\cdot 4$, $b=-1\cdot 2$
   (g) $s=3$, $t=-4$ (h) $a=0$, $b=4$ (i) $x=1$, $y=-3$ (j) $a=-2$, $b=3$

5 (a) max value: 12; min value: 0 (b) $x\geqslant 0$, $y\geqslant 0$, $x+3y\leqslant 12$

6 (a) max value: 12; min value: 0 (b) $x\geqslant 0$; $y\geqslant 0$; $2x+3y\leqslant 18$

7 (a) $x\geqslant 0$; $y\geqslant 0$; $2x+y\leqslant 12$; $x+y\leqslant 10$ (b) max value: 12; min value: 0

8 (a)

(b)

(c)

(d)

10 (a)

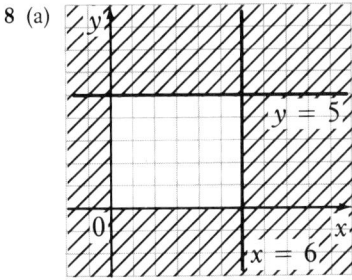

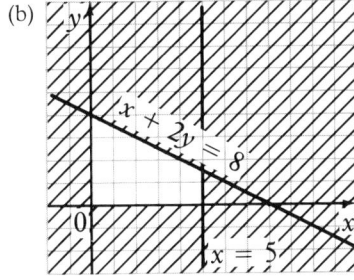

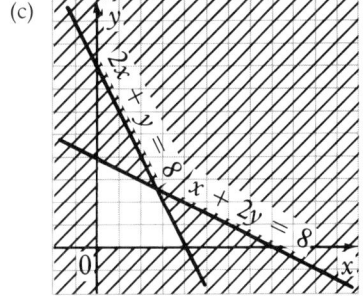

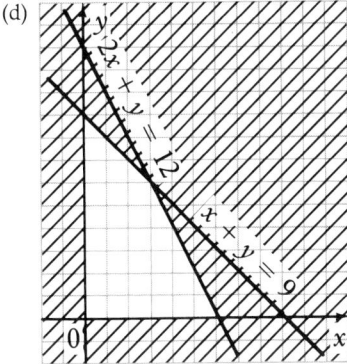

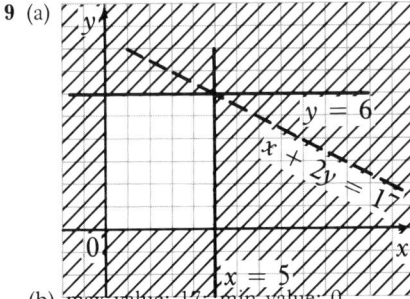

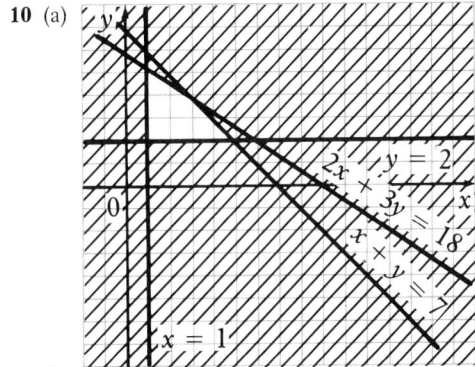

(b) max value: 22 at point (5,2)   min value: 6 at point (1,2)

9 (a)

(b) max value: 17; min value: 0

11 (a) Let x be number of dogs. Let y be number of cats. $x \geqslant 0$; $y \geqslant 0$; $x+y \leqslant 60$; $20x+12y \leqslant 1000$

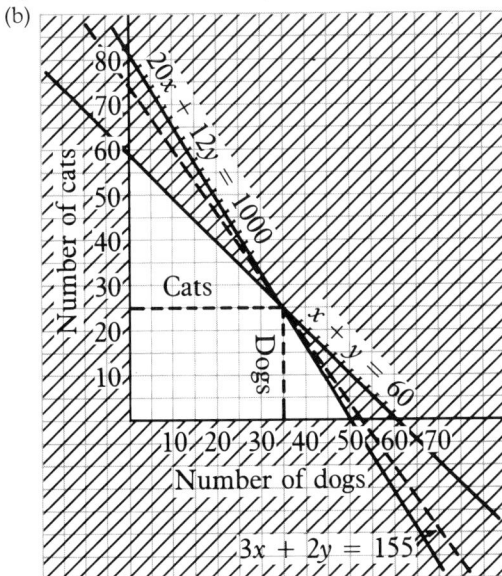

(b)

(c) Spaces for dogs = 35. Spaces for cats = 25   (d) £155

(e) The likely demand from pet owners. Would there be more demand from dog owners or cat owners?
The costs of feeding and looking after the different animals.

12 Burgers = 115 kg; bangers = 95 kg

Revision—Simultaneous equations and linear programming

1 (a) $x=3$; $y=2$ (b) $x=2$; $y=-1$ (c) $a=-2$; $b=3$ (d) $x=-1$; $y=-2$ (e) $a=1\frac{1}{2}$; $b=-2\frac{1}{2}$
(f) $x=-2$; $y=-3$

2 (a) $x\geqslant0$; $y\geqslant0$; $x\leqslant6$; $y\leqslant4$ (b) max value: 10; min value: 0

3 (a) $x\geqslant0$; $y\geqslant0$; $2x+y\leqslant10$; $2x+3y\leqslant12$ (b) max value: 10; min value: 0

4 (a)

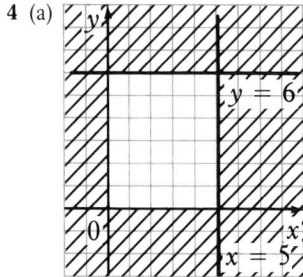

(b)

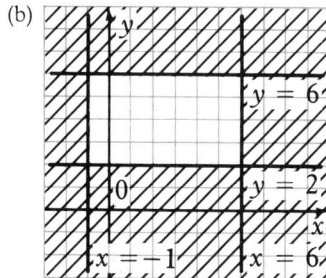

(c)

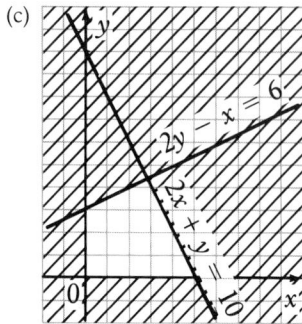

(d)

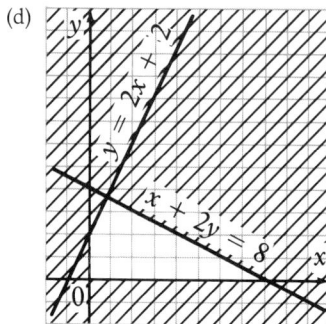

Unit 4. Indices and algebraic fractions

1 (a) 3^7 (b) 2^{12} (c) $2^3\times3^4$ (no change) (d) 5^{3+x} (e) a^7 (f) a^{m+n} (add the indices)

2 5^3 (b) 3^2 (c) $5^4\div2^3$ (no change) (d) a^4 (e) a^{m-n} (subtract the indices)
(f) 3^{4-x} Index $\leqslant0$ when $x\geqslant4$.

3 (a) x^7 (b) b^4 (c) x^3 (d) a (e) $2x^5$ (f) $6a^7$ (g) $3x^2$ (h) $4m^6$ (i) $6x^3y^2$ (j) 3^5 (k) 5^3 (l) x^2
(m) a^4 (n) 2 (o) $2p^2+6p$

4 (a) $\frac{1}{4}$ (b) 1 (c) 25 (d) 16 (e) 9 (f) $\frac{1}{9}$

5 (a) x^8 (b) a^8 (c) b^2 (d) x^7 (e) a (f) x^3 (g) $4x^2$ (h) a^6b^3 (i) a^5 (j) 1 (k) $2x$ (l) x^{-8}
(m) $6a^3b^3$ (n) $x^{-1}+x^4$

6 (a) 3 (b) 3 (c) 6 (d) 2 (e) $\frac{1}{7}$ (f) 8 (g) 27 (h) 9 (i) 4 (j) 5 (k) 20 (l) 20

7 (a) 16 (b) 1 (c) -48 (d) 2 **8** (a) 2^2 (b) 2^{-2} (c) 2^6 (d) $2^{5/2}$ **9** 3^{21}

10 (a) 3×10^9 (b) $2\cdot4\times10^9$ (c) 3×10^8 (d) 1×10^{-9} **11** (a) 8×10^5 (b) 2×10^7 (c) $5\cdot6\times10^3$ (d) $4\cdot8\times10^{-7}$

12 (b) and (d) **13** (a) $2\sqrt{2}$ (b) $4\sqrt{2}$ (c) $3\sqrt{3}$ (d) $2\sqrt{5}$ (e) $\frac{2}{3}\sqrt{2}$ (f) $\frac{3}{5}\sqrt{5}$ (g) $10\sqrt{3}$ (h) $2\sqrt[3]{2}$ (i) $2\sqrt[3]{-2}$

14 (a) 3 (b) $2\sqrt{3}$ (c) 5 (d) 6 (e) 20 (f) 9 **15** (a) $2\sqrt{3}$ (b) 10 (c) $10\sqrt{10}$ (d) 30 (e) 30 (f) $5\sqrt{3}$

16 (a) $\frac{1}{5}\sqrt{5}$ (b) $\sqrt{2}$ (c) $\frac{1}{2}\sqrt{6}$ (d) $\frac{1}{10}\sqrt{3}$ (e) $\sqrt{7}$ (f) $\frac{1}{2}\sqrt{2}$ **17** (a) $\frac{1}{2}$ (b) $\sqrt{3}$ (c) $\frac{1}{2}$ (d) $\frac{1}{\sqrt{3}}=\frac{\sqrt{3}}{2}$ (e) $\frac{\sqrt{3}}{2}$

18 (a) $\frac{xa}{y}$ (b) $\frac{x^2}{2y}$ (c) $\frac{ac}{b^2}$ (d) $\frac{2x^3}{y^2}$ (e) $\frac{b^2c}{a}$ (f) $\frac{ac}{bd-be}$ **19** (a) $\frac{ps}{qr}$ (b) $\frac{x^2}{2y}$ (c) $\frac{b^3}{ac}$ (d) $\frac{2x^3}{y^2}$ (e) $\frac{x}{ay}$ (f) $\frac{ab^2}{c}$

20 (a) $\frac{b}{c}$ (b) $\frac{2}{x}$ (c) $\frac{3}{2}b$ (d) $\frac{5}{3}pq$ (e) $\frac{b}{x+y}$ (f) $\frac{n+2}{3}$

21 (a) $\frac{b+a}{ab}$ (b) $\frac{a^2+2}{2a}$ (c) $\frac{bx-ay}{by}$ (d) $\frac{x^2-4}{2x}$ (e) $\frac{a^2-2}{4ab}$ (f) $\frac{3x+2}{x(x+1)}$ (g) $\frac{2y}{a(a+2)}$ (h) $\frac{1}{a^2(a^2+1)}$ (i) $\frac{6x}{(x-3)(x+3)}$

22 (a) $x=8$ (b) $x=-2$ (c) $x=\frac{1}{2}$ (d) $x=6$ (e) $x=-1$ (f) $x=4$ (g) $x=\pm6$ (h) $a=3\sqrt{2}$ (i) $x=-4$

Revision—Indices and algebraic fractions

1 (a) p^7 (b) $\frac{2}{3}x^3$ (c) $12a^5$ (d) $3b^2$ (e) 2^5 (f) $\frac{2}{x}$ **2** (a) $\frac{1}{9}$ (b) 1 (c) 2 (d) 5^2 (e) 8 (f) $\frac{1}{6}$

3 (a) $6a^3b^3$ (b) $1+a^5$ (c) $\frac{x}{2y}$ (d) $3p^2$ **4** (a) 6×10^6 (b) $2\cdot5\times10^9$

5 (a) $1\cdot6\times10^8$ (b) 5×10^{-2} **6** (a) $3\sqrt2$ (b) $10\sqrt3$ (c) $6\sqrt2$ (d) $5\sqrt3$

7 (a) $\frac{\sqrt7}{7}$ (b) $\frac{2\sqrt6}{3}$ (c) $2\sqrt2$ **8** (a) $\frac{2a^3}{b^2}$ (b) $\frac{a^2}{2b}$ (c) a (d) $\frac{p}{q}$

9 (a) $\frac{y+x}{xy}$ (b) $\frac{a^2-8}{2ab}$ (c) $\frac{2}{x^2(x^2-2)}$ (d) $\frac{-x-25}{(x+5)(x-5)}$ **10** (a) $x=-8$ (b) $x=3$

Unit 5. Quadratics

1

expression	factors	zeros	minimum	minimum value
x^2+x-2	$(x+2),\ (x-1)$	$x=-2,\ x=1$	$x=-0\cdot5$	$-2\cdot25$
x^2-x-6	$(x-3),\ (x+2)$	$x=3,\ x=-2$	$x=0\cdot5$	$-6\cdot25$
x^2-4x+3	$(x-3),\ (x-1)$	$x=3,\ x=1$	$x=2$	-1
x^2+5x+6	$(x+3),\ (x+2)$	$x=-3,\ x=-2$	$x=-2\cdot5$	$-0\cdot25$
$x^2-3x-10$	$(x-5),\ (x+2)$	$x=5,\ x=-2$	$x=1\cdot5$	$-12\cdot25$
$x^2-5x-24$	$(x-8),\ (x+3)$	$x=8,\ x=-3$	$x=2\cdot5$	$-30\cdot25$

2

expression	zeros	maximum	maximum value
(a) $3-2x-x^2$	$x=1,\ x=-3$	$x=-1$	4
(b) $3+2x-x^2$	$x=-1,\ x=3$	$x=1$	4
(c) $6-x-x^2$	$x=-3,\ x=2$	$x=-0\cdot5$	$6\cdot25$

3 (a) $x^2-2x-15=(x-5)(x+3)$; the roots are $x=5$ and $x=-3$; the **minimum** value is -16, when $x=1$.
  (b) $25-5x-2x^2=(5+x)(5-2x)$; the roots are $x=-5$ and $x=2\cdot5$; the **maximum** value is $28\cdot125$, when $x=-1\cdot25$.
  (c) $12x^2-7x-10=(3x+2)(4x-5)$; the roots are $x=-\frac{2}{3}\ (-0\cdot666\ldots)$ and $x=\frac{5}{4}\ (1\cdot25)$; the **minimum** value is $-11\cdot02\ldots$, when $x=0\cdot2917\ldots$.
  (d) $x^2+5x=x(x+5)$; the roots are $x=0$ and $x=-5$; the **minimum** value is $-6\cdot25$, when $x=-2\cdot5$.
  (e) $x-x^2=x(1-x)$; the roots are $x=0$ and $x=1$; the **maximum** value is $\frac{1}{4}$ when $x=\frac{1}{2}$.

4 (a) the sides of the rectangle are 9 metres and 1 metre
  (b) the sides of the rectangle are 7 metres and 3 metres
  (c) the sides of the rectangle are 4 metres and 1 metre
  (d) the sides of the rectangle (square) are 10 metres and 10 metres

5 (a) $x=-4,\ x=3$ (b) $x=-2,\ x=5$ (c) $x=-1,\ x=5$ (d) $x=7,\ x=2$ (e) $x=2,\ x=-\frac{3}{2}$ (f) $x=-6,\ x=3$

6 $x=9\cdot36$ (to 2 decimal places)

7 and **8** (a) $x=8\cdot16$ and $x=1\cdot84$ (b) $x=10\cdot525$ and $x=0\cdot475$ (c) $x=6\cdot46$ and $x=-0\cdot46$

9 (a) $x^2+x-13=0$; $a=1,\ b=1,\ c=-13$ (b) $x^2-3x-9=0$; $a=1,\ b=-3,\ c=-9$
  (c) $-x^2+4x+7=0$; $a=-1,\ b=4,\ c=7$ (d) $-x^2+9x-13=0$; $a=-1,\ b=9,\ c=-13$
  (e) $-x^2-x+24=0$; $a=-1,\ b=-1,\ c=24$ (f) $2x^2-9x+10=0$; $a=2,\ b=-9,\ c=10$

10 (a) $x=3\cdot14$ and $x=-4\cdot14$ (b) $x=4\cdot85$ and $x=-1\cdot85$ (c) $x=-1\cdot32$ and $x=5\cdot32$ (d) $x=1\cdot81$ and $x=7\cdot19$
  (e) $x=-5\cdot42$ and $x=4\cdot42$ (f) $x=2\cdot5$ and $x=2$

11 (a) Let $x=$ the length of the rectangle, then the width of the rectangle is $20-x$.
    $x(20-x)=90$ $x=13\cdot2$ and $x=6\cdot8$. The length of the rectangle is $13\cdot2$ cm and the width is $6\cdot8$ cm
  (b) $n^2=121$; $n=11$. The first 11 odd numbers add up to 121.
  (c) $n(n+1)=210$; $n=14$ and $n=-15$. The first 14 even numbers add up to 210.
  (d) Let the length of the shortest side be x; then the length of the hypotenuse is $x+3$, and the length of the other side is $x+1$.
    $(x+3)^2=x^2+(x+1)^2$; $x=5\cdot46$ cm and $x=-1\cdot46$ cm (We cannot use $x=-1\cdot46$ for the length of a side.)
    The length of shortest side is $5\cdot46$ cm, the length of the hypotenuse is $8\cdot46$ cm and the length of the other side is $6\cdot46$ cm.

12 (a) $x=7$ and $x=1$ (b) $x=2\cdot41$ and $x=-0\cdot41$ (c) $x=1\cdot73$ and $x=-1\cdot73$

Revision—Quadratics

1 (a) $x^2+6x+8=(x+2)(x+4)$; $x=-2$, $x=-4$ minimum value -1 when $x=-3$

(b) $x^2-6x+5=(x-5)(x-1)$; $x=5$, $x=1$ minimum value -4 when $x=3$

(c) $x^2-4x-5=(x+1)(x-5)$; $x=-1$, $x=5$ minimum value -9 when $x=2$

(d) $x^2+5x-6=(x-1)(x+6)$; $x=1$, $x=-6$ minimum value $-12\cdot25$ when $x=-2\cdot5$

2 (a) $28-3x-x^2=(7+x)(4-x)$; $x=-7$, $x=4$ maximum value $30\cdot25$ when $x=-1\cdot5$

(b) $18+3x-x^2=(3+x)(6-x)$; $x=-3$, $x=6$ maximum value $20\cdot25$ when $x=1\cdot5$

(c) $28-27x-x^2=(28+x)(1-x)$; $x=-28$, $x=1$ maximum value $210\cdot25$ when $x=-13\cdot5$

(d) $18+17x-x^2=(1+x)(18-x)$; $x=-1$, $x=18$ maximum value $90\cdot25$ when $x=8\cdot5$

3 (a) maximum value $56\cdot3$ when $x=-0\cdot67$ (b) minimum value -4 when $x=-4$

(c) maximum value $40\cdot5$ when $x=1\cdot5$ (d) minimum value $-4\cdot27$ when $x=-0\cdot13$

(e) minimum value -9 when $x=3$ (f) maximum value 1 when $x=1$

4 (a) $x=6$ and $x=-5$ (b) $x=1$ and $x=-4$ (c) $x=-1$ and $x=10$ (d) $x=2$ and $x=6$ (e) $x=\frac{5}{3}$ $(1\cdot67)$ and $x=-1$

(f) $x=8$ and $x=-\frac{1}{2}$ $(-0\cdot5)$

5 $x=7\cdot16$ and $x=0\cdot838$

6 (a) $9\cdot27$, $1\cdot73$ (b) $8\cdot41$, $0\cdot59$ (c) $7\cdot27$, $-0\cdot27$ (d) $-5\cdot54$, $0\cdot54$ (e) $4\cdot32$, $-2\cdot32$

(f) $1\cdot46$, $-5\cdot46$ (g) $-1\cdot37$, $4\cdot37$ (h) $1\cdot76$, $6\cdot24$ (i) $-2\cdot76$, $3\cdot26$ (j) -3, $-\frac{4}{3}$ $(-1\cdot33)$

7 (a) $x=1$, $x=-2\cdot57$ (b) $x=4$, $x=-1\cdot8$ (c) $x=1\cdot56$, $x=-2\cdot56$

8 (a) Let $x=$ length of base, then height $=x+10$; $\frac{1}{2}x(x+10)=50$. Length of base $=6\cdot18$ cm, height $=16\cdot18$ cm

(b) $\frac{1}{2}n(n+1)=465$, so $n=30$. The first 30 whole numbers are needed to add up to 465.

(c) Let the length of the hypotenuse be x, then the length of one side is $x-5$ and the length of the other side is $\frac{1}{2}x$

$x^2=(x-5)^2+(\frac{1}{2}x)^2$; $x=37\cdot32$ and $x=2\cdot68$ (we cannot use $2\cdot68$ since one side cannot be '5 less than $2\cdot68$')

The length of the hypotenuse is $37\cdot32$ cm; the length of one side is $32\cdot32$ cm; the length of the other side is $18\cdot66$ cm

(d) The numbers are 0 and 100 or $33\cdot33$ and $133\cdot33$

Unit 6. Functions

1

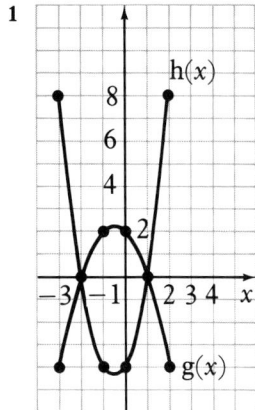

2 (see next page)

3

x	x^3-2x^2-5x+6	$(x,f(x))$
4	18	(4,18)
3	0	(3,0)
2	-4	$(2,-4)$
1	0	(1,0)
0	6	(0,6)
-1	8	$(-1,8)$
-2	0	$(-2,0)$
-3	-24	$(-3,-24)$

Roots of equation $x^3-2x^2-5x+6=0$ are $x=-2$, $x=1$, $x=3$

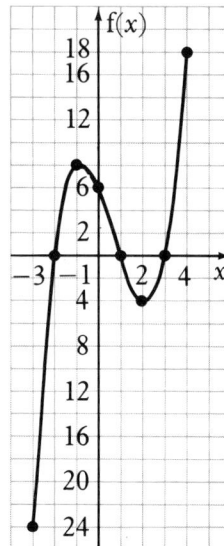

4

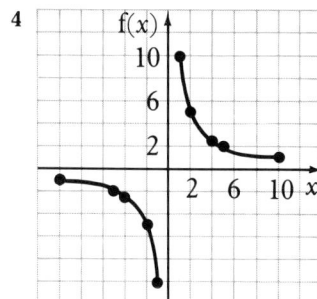

2

interval		number line
$-1 < x < 3$	$x \in R$	
$1 \leqslant x \leqslant 4$	$x \in R$	
$-4 < x \leqslant 0$	$x \in R$	
$-2 \leqslant x < 2$	$x \in R$	
$-5 \leqslant x \leqslant -1$	$x \in R$	
$0 < x < 4$	$x \in R$	

5 (a)

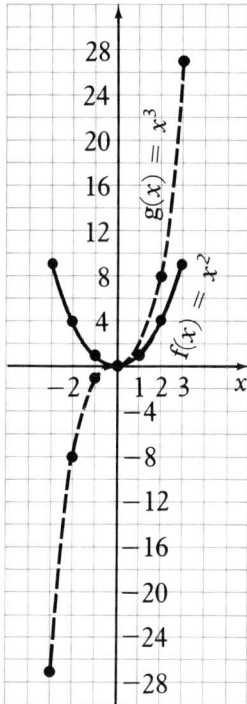

5 (b) $g(x) = x^3$ is the odd function

(c) The graph of an even function is symmetrical in the f(x)-axis (the y-axis), i.e. if a point on the graph is reflected in the y-axis, the image of the point is also on the graph.

If we change the sign of a value of x, the value of f(x) does not change

6

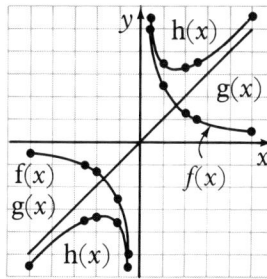

7 (a)

n	3^n	$(n, f(n))$
1	3	(1,3)
2	9	(2,9)
3	27	(3,27)
4	81	(4,81)

7 (b)

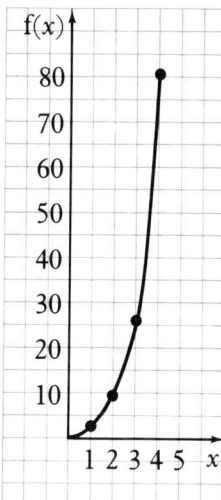

8

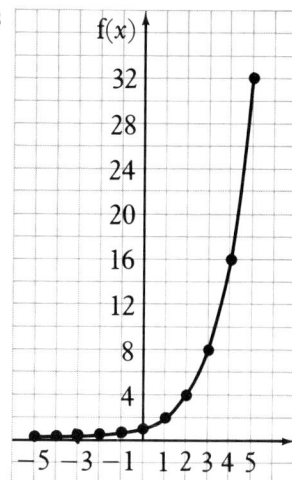

9

x	100×3^{-x}	$(x, h(x))$
0	$100 \times 3^0 = 100 \times 1 = 100$	$(0, 100)$
1	$100 \times 3^{-1} = 100 \times \frac{1}{3} = 33$	$(1, 33)$
2	$100 \times 3^{-2} = 100 \times \frac{1}{9} = 11$	$(2, 11)$
3	$100 \times 3^{-3} = 100 \times \frac{1}{27} = 4$	$(3, 4)$
4	$100 \times 3^{-4} = 100 \times \frac{1}{81}$	$(4, 1)$

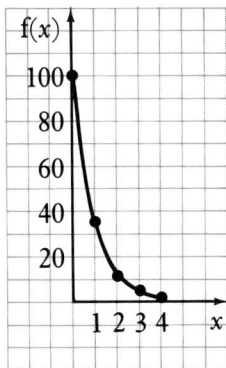

10 $\dfrac{y+4}{5} \xleftarrow[\text{divide by 5}]{\div 5} y+4 \xleftarrow[\text{add 4}]{+4} y$ $x = f^{-1}(y) = \dfrac{y+4}{5}$ $y \in R$

11 (a) $f^{-1}(y) = \dfrac{y-7}{2}$ (b) $g^{-1}(y) = 2(y+6)$ (c) $h^{-1}(y) = 3y - 2$ **12** $f^{-1}(y) = -\sqrt{y}, \ y \geqslant 0, \ y \in R$

13 $\sqrt{\dfrac{y}{3}} \xleftarrow[\text{square root}]{\sqrt{\ }} \dfrac{y}{3} \xleftarrow[\text{divide by 3}]{\div 3} y$ $x = f^{-1}(y) = \sqrt{\dfrac{y}{3}}$ $y \in R$

14 (a) $f^{-1}(y) = \sqrt{2x}; \ y \geqslant 0 \ y \in R$ (b) $g^{-1}(y) = \sqrt{(y+1)}; \ y \geqslant -1, \ y \in R$
   (c) $h^{-1}(y) = \sqrt{2(y+1)}; \ y \geqslant -1, \ y \in R$

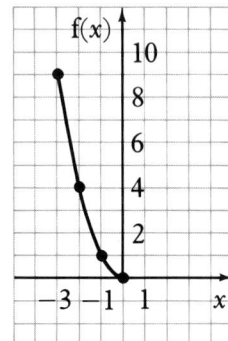

Revision—Functions

1 (a)

x	$x^2 - x - 2$	$(x, f(x))$
4	10	$(4, 10)$
3	4	$(3, 4)$
2	0	$(2, 0)$
1	-2	$(1, -2)$
0	-2	$(0, -2)$
-1	0	$(-1, 0)$
-2	4	$(-2, 4)$
-3	10	$(-3, 10)$

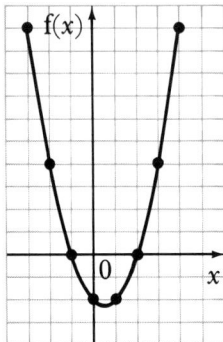

(c)

x	$2x^2 + x - 6$	$(x, h(x))$
3	15	$(3, 15)$
2	4	$(2, 4)$
1	-3	$(1, -3)$
0	-6	$(0, -6)$
-1	-5	$(-1, -5)$
-2	0	$(-2, 0)$
-3	9	$(-3, 9)$

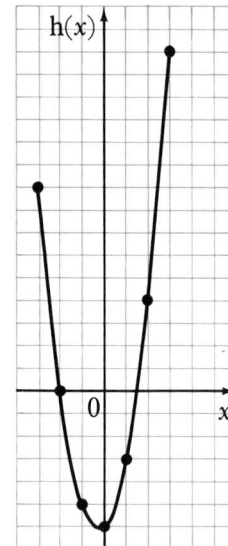

(b)

x	$6 - x - x^2$	$(x, g(x))$
3	-6	$(3, -6)$
2	0	$(2, 0)$
1	4	$(1, 4)$
0	6	$(0, 6)$
-1	6	$(-1, 6)$
-2	4	$(-2, 4)$
-3	0	$(-3, 0)$
-4	-6	$(-4, -6)$

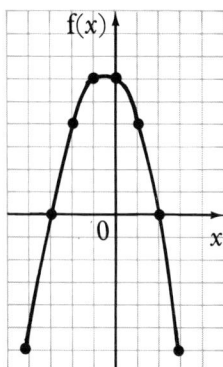

2 (a)

x	$(x+2)(x-1)(x-3)$	$(x,j(x))$
4	18	(4,18)
3	0	(3,0)
2	-4	$(2,-4)$
1	0	(1,0)
0	6	(0,6)
-1	8	$(-1,8)$
-2	0	$(-2,0)$
-3	-24	$(-3,-24)$

(b)

x	x^3+2x^2-x-2	$(x,k(x))$
3	40	(3,40)
2	12	(2,12)
1	0	(1,0)
0	-2	$(0,-2)$
-1	0	$(-1,0)$
-2	0	$(-2,0)$
-3	-8	$(-3,-8)$

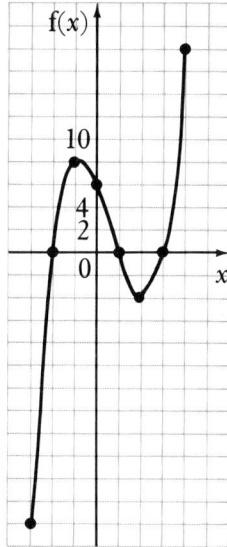

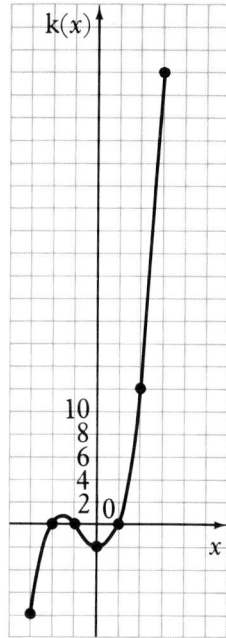

3 (a) $f(1)=1$, $f(-1)=-1$, $f(x)$ is odd (b) $f(1)=0$, $f(-1)=-2$, $f(x)$ is neither even nor odd
(c) $f(1)=1$, $f(-1)=-1$, $f(x)$ is odd (d) $f(1)=1$, $f(-1)=1$, $f(x)$ is even
(e) $f(1)=16$, $f(-1)=4$, $f(x)$ is neither even nor odd (f) $f(1)=1$, $f(-1)=1$, $f(x)$ is even
(g) $f(1)=2$, $f(-1)=\frac{1}{2}$, $f(x)$ is neither even nor odd (h) $f(1)=1$, $f(-1)=-1$, $f(x)$ is odd
(i) $f(1)=-1$, $f(-1)=-27$, $f(x)$ is neither even nor odd (j) $f(1)=1$, $f(-1)=-1$, $f(x)$ is odd
(k) $f(1)=3$, $f(-1)=\frac{1}{3}$, $f(x)$ is neither even nor odd

4 (a)

n	$(1\cdot5)^n$	$(n,f(n))$
4	$5\cdot1$	$(4,5\cdot1)$
3	$3\cdot4$	$(3,3\cdot4)$
2	$2\cdot25$	$(2,2\cdot25)$
1	$1\cdot5$	$(1,1\cdot5)$
0	1	(0,1)
-1	$0\cdot7$	$(-1,0\cdot7)$
-2	$0\cdot4$	$(-2,0\cdot4)$
-3	$0\cdot3$	$(-3,0\cdot3)$
-4	$0\cdot2$	$(-4,0\cdot2)$

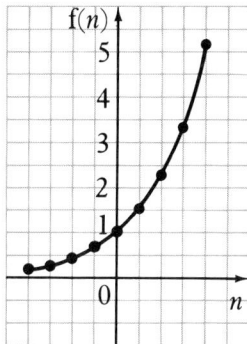

5 (a) $x \xrightarrow[\text{multiply by 3}]{\times 3} 3x \xrightarrow[\text{subtract 2}]{-2} 3x-2$ (b) $x \xrightarrow[\text{divide by 4}]{\div 4} \frac{1}{4}x \xrightarrow[\text{add 3}]{+3} \frac{1}{4}x+3$

(c) $x \xrightarrow[\text{divide by 6}]{\div 6} \frac{x}{6} \xrightarrow[\text{subtract 5}]{-5} \frac{x}{6}-5$ (d) $x \xrightarrow[\text{subtract 3}]{-3} x-3 \xrightarrow[\text{divide by 2}]{\div 2} \frac{x-3}{2}$

(e) $x \xrightarrow[\text{add 2}]{+2} x+2 \xrightarrow[\text{divide by 3}]{\div 3} \frac{1}{3}(x+2)$ (f) $x \xrightarrow[\text{square}]{(\ )^2} x^2 \xrightarrow[\text{add 2}]{+2} x^2+2$ (g) $x \xrightarrow[\text{square}]{(\ )^2} x^2 \xrightarrow[\text{multiply by 3}]{\times 3} 3x^2$

(h) $x \xrightarrow[\text{square}]{(\ )^2} x^2 \xrightarrow[\text{multiply by 3}]{\times 3} 3x^2 \xrightarrow[\text{add 2}]{+2} 3x^2+2$

6 (b) $y=g(x)=2x-3 \quad x\in R$

$$x \xrightarrow[\text{multiply by 2}]{\times 2} 2x \xrightarrow[\text{subtract 3}]{-3} 2x-3$$

$$\frac{y+3}{2} \xleftarrow[\text{divide by 2}]{\div 2} y+3 \xleftarrow[\text{add 3}]{+3} y$$

$$x=f^{-1}(y)=\frac{y+3}{2}$$

7 (a) $x = f^{-1}(y) = \dfrac{y+1}{2}$ $y \in R$ (b) $x = g^{-1}(y) = \dfrac{y}{2} + 1$ $y \in R$ (c) $x = h^{-1}(y) = 3(y+4)$ $y \in R$

(d) $x = j^{-1}(y) = 5(y-6)$ $y \in R$ (e) $x = k^{-1}(y) = 3y - 2$ $y \in R$ (f) $x = m^{-1}(y) = 2y + 3$ $y \in R$

(g) $x = n^{-1}(y) = \sqrt{(y+3)}$ $y \geqslant -3,\ y \in R$ (h) $x = p^{-1}(y) = \sqrt{\dfrac{y}{4}}$ $y \geqslant 0,\ y \in R$ (i) $x = q^{-1}(y) = \sqrt{\dfrac{y+3}{4}}$ $y \geqslant -3,\ y \in R$

Unit 7.   Trig functions and graphs

1

| 0 sec | ²⁄₁₀ sec | ⁴⁄₁₀ sec | ⁶⁄₁₀ sec | ⁸⁄₁₀ sec | 1 sec |

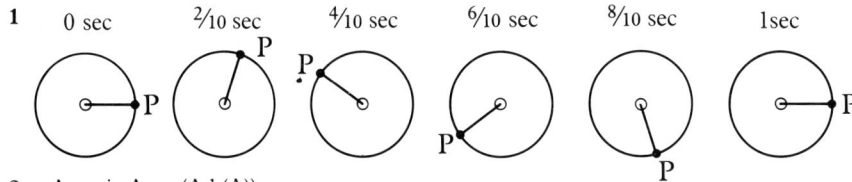

2

A	sin A	(A,h(A))
0°	0	(0,0)
36°	0·59	(36°,0·59)
72°	0·95	(72°,0·95)
108°	0·95	(108°,0·95)
144°	0·59	(144°,0·59)
180°	0	(180°,0)
216°	−0·59	(216°,−0·59)
252°	−0·95	(252°,−0·95)
288°	−0·95	(288°,−0·95)
324°	−0·59	(324°,−0·59)
360°	0	(360°,0)

3 (a)

$\sin 50° = 0·77$
$\sin 130° = \sin 50°$ $\sin 130° = 0·77$
$\sin 230° = -\sin 50°$ $\sin 230° = -0·77$
$\sin 310° = -\sin 50°$ $\sin 310° = -0·77$

(b)

$\sin 70° = 0·940$
$\sin 110° = \sin 70°$ $\sin 110° = 0·940$
$\sin 250° = -\sin 70°$ $\sin 250° = -0·940$
$\sin 290° = -\sin 70°$ $\sin 290° = -0·940$

4 (a)

A	sin A
396°	0·59
432°	0·95
468°	0·95
504°	0·59
540°	0
576°	−0·59
612°	−0·95
648°	−0·95
684°	−0·59
720°	0

(b)

A	sin A
−36°	−0·59
−72°	−0·95
−108°	−0·95
−144°	−0·59
−180°	0
−216°	0·59
−252°	0·95
−288°	0·95
−324°	0·59
−360°	0

6

A	cos A	(A,d(A))
0°	1	(0°,1)
36°	0·81	(36°,0·81)
72°	0·31	(72°,0·31)
108°	−0·31	(108°,−0·31)
144°	−0·81	(144°,−0·81)
180°	−1	(180°,−1)
216°	−0·81	(216°,−0·81)
252°	−0·31	(252°,−0·31)
288°	0·31	(288°,0·31)
324°	0·81	(324°,0·81)
360°	1	(360°,1)

5

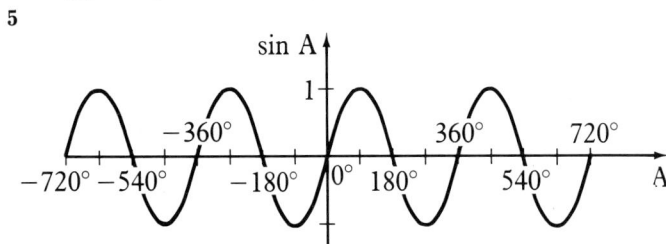

7 (a)

$\cos 70° = 0·34$
$\cos 110° = -\cos 70°$ $\cos 110° = -0·34$
$\cos 250° = -\cos 70°$ $\cos 250° = -0·34$
$\cos 290° = \cos 70°$ $\cos 290° = 0·34$

(b)

$\cos 50° = 0·643$
$\cos 130° = -\cos 50°$ $\cos 130° = -0·643$
$\cos 230° = -\cos 50°$ $\cos 230° = -0·643$
$\cos 310° = \cos 50°$ $\cos 310° = 0·643$

8

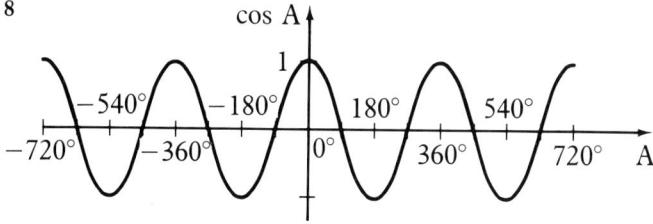

9 (a)

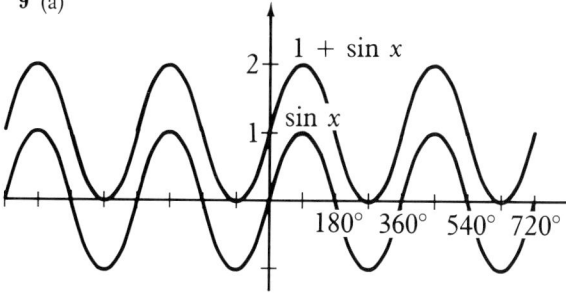

(b)

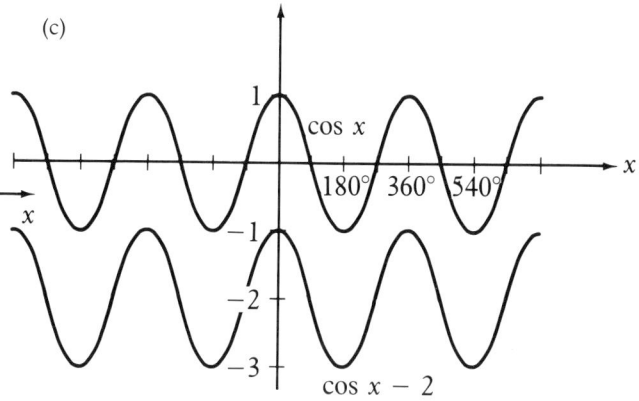

(c)

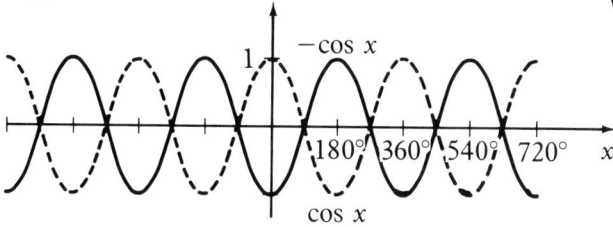

(d)

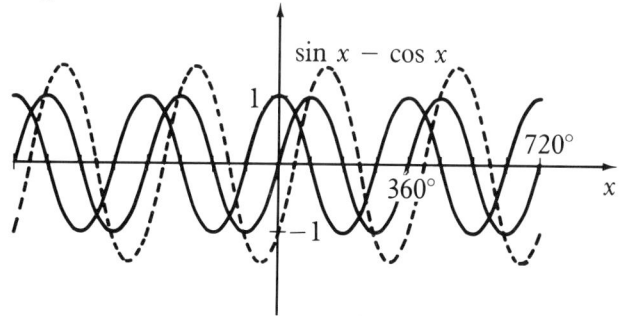

10 (a)

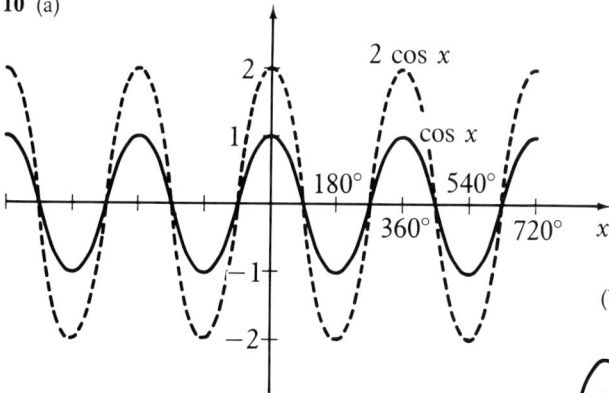

(b)

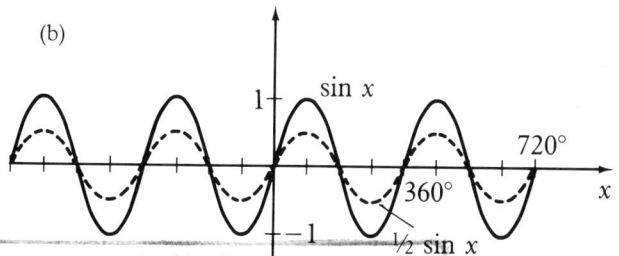

11 (a) period of $\sin 2x = 180°$; amplitude $= 1$

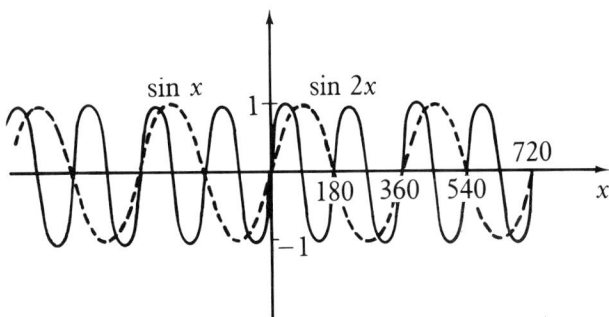

(b) period of $\cos 3x = 120°$; amplitude $= 1$

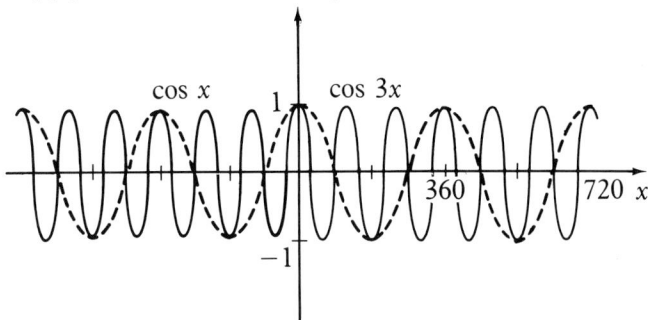

(c) period of $\frac{1}{2}\sin 3x = 120°$; amplitude $= \frac{1}{2}$

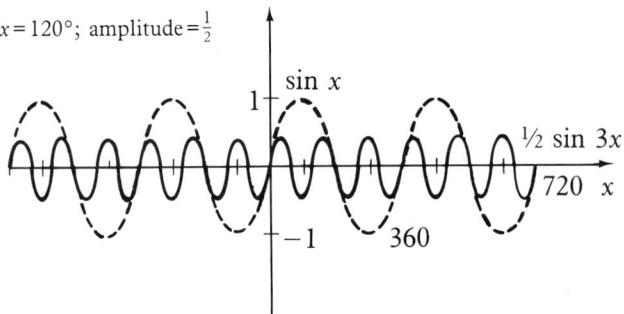

12 (a) period of $\cos \frac{1}{2}x = 720°$; amplitude $= 1$

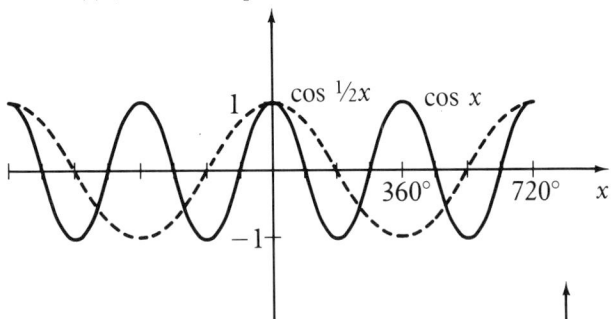

(b) period of $\sin \frac{1}{3}x = 1080°$ $(3 \times 360°)$; amplitude $= 1$

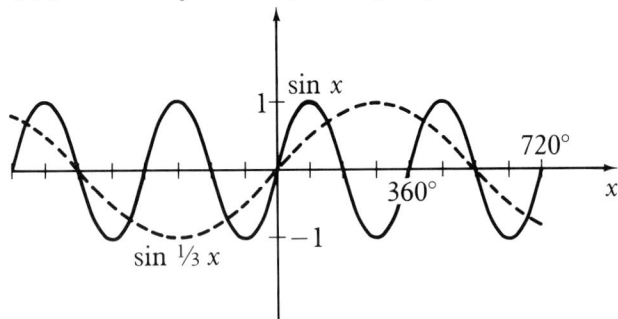

(c) period of $4\cos \frac{1}{3}x = 1080°$; amplitude $= 4$

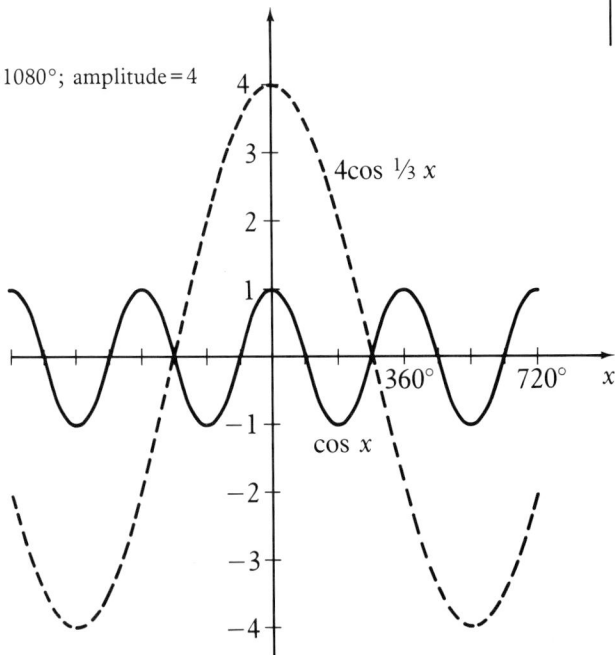

13 (a)

sin (x + 90°)

720°

360°

x

1

−1

sin x

(b)

cos x

720°

360°

x

1

−1

cos (x − 90°)

(c)

sin (x + 180°)

1

360°   720°   x

−1

sin x

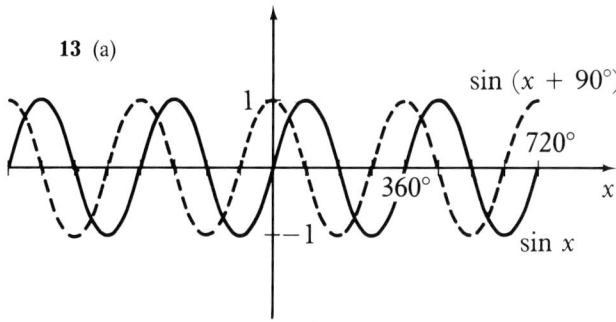

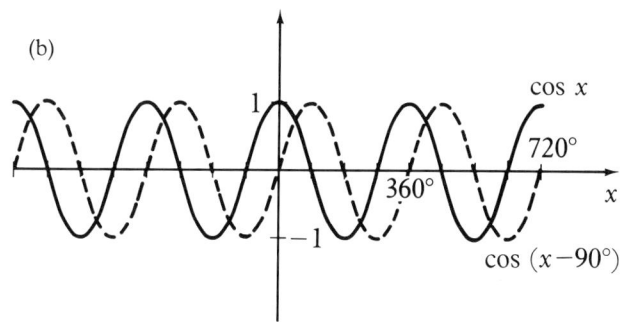

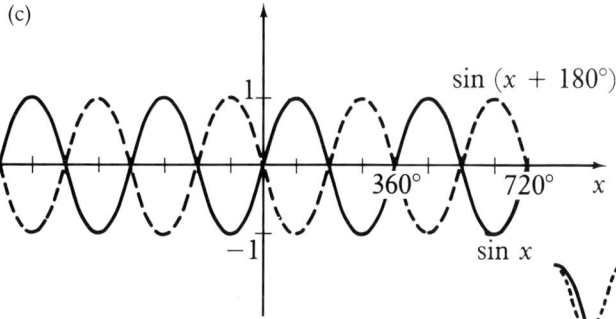

14 Period of $\cos^2 x = 180°$; amplitude $= 1$.

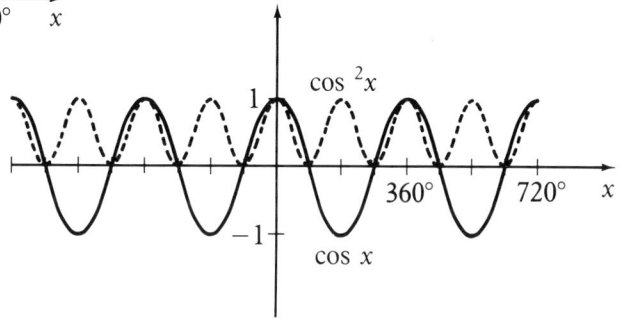

$\cos^2 x$

1

360°   720°   x

−1

cos x

15 (a) $3\sin\frac{1}{2}x$; $a = 3$, $b = \frac{1}{2}$ (b) $2\cos(x + 90°)$; $a = 2$, $c = 90°$
  (c) $\frac{1}{2}\cos 3x$; $a = \frac{1}{2}$, $b = 3$ (d) $\frac{1}{2}\sin(x + 45°)$; $a = \frac{1}{2}$, $c = 45°$

16

A	0°	20°	40°	60°	80°	100°	120°	140°	160°	180°
tan A	0	0·36	0·84	1·73	5·67	−5·67	−1·73	−0·84	−0·36	0

A	200°	220°	240°	260°	280°	300°	320°	340°	360°
tan A	0·36	0·84	1·73	5·67	−5·67	−1·73	−0·84	−0·36	0

17 (a)

tan 30° = 0·58
tan 150° = −tan 30°   tan 150° = −0·58
tan 210° = tan 30°   tan 210° = 0·58
tan 330° = −tan 30°   tan 330° = −0·58

(b)

tan 60° = 1·73
tan 120° = −tan 60°   tan 120° = −1·73
tan 240° = tan 60°   tan 240 = 1·73
tan 300° = −tan 60°   tan 300° = −1·73

18

tan A

4

3

2

1

−360°   0°   90°   180°   270°   360° A

−1

−2

−3

−4

19

A	15°	54°	57°	123°	234°	345°
sin A	0·26	0·81	0·84	0·84	−0·81	−0·26
cos A	0·97	0·59	0·54	−0·54	−0·59	0·97
tan A	0·27	1·38	1·54	−1·54	1·38	−0·27

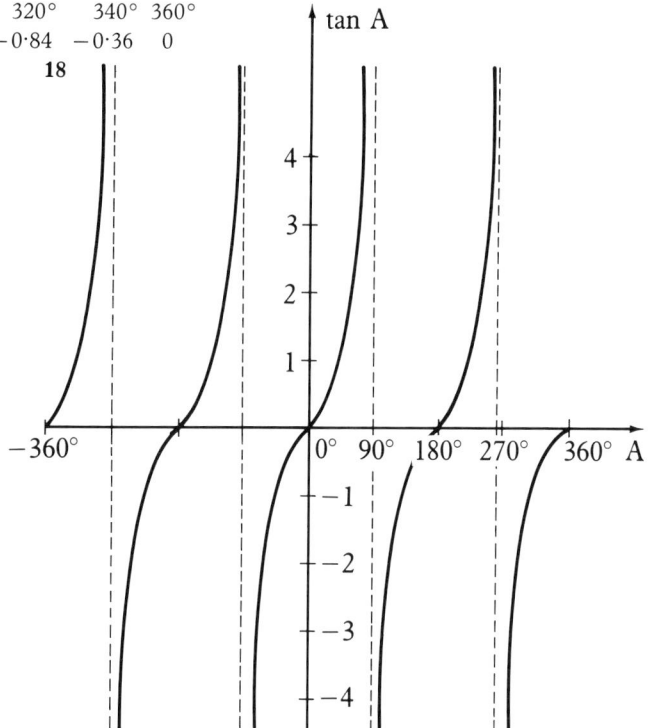

Revision—Trig functions and graphs

1 (a)

	$\sin 33° = 0·545$	(b)		$\sin 77° = 0·974$
$\sin 147° = \sin 33°$	$\sin 147° = 0·545$		$\sin 103° = \sin 77°$	$\sin 103° = 0·974$
$\sin 213° = -\sin 33°$	$\sin 213° = -0·545$		$\sin 257° = -\sin 77°$	$\sin 257° = -0·974$
$\sin 327° = -\sin 33°$	$\sin 327° = -0·545$		$\sin 283° = -\sin 77°$	$\sin 283° = -0·974$

2 (a)

	$\cos 44° = 0·719$	(b)		$\cos 66° = 0·407$
$\cos 136° = -\cos 44°$	$\cos 136° = -0·719$		$\cos 114° = -\cos 66°$	$\cos 114° = -0·407$
$\cos 224° = -\cos 44°$	$\cos 224° = -0·719$		$\cos 246° = -\cos 66°$	$\cos 246° = -0·407$
$\cos 316° = \cos 44°$	$\cos 316° = 0·719$		$\cos 294° = \cos 66°$	$\cos 294° = 0·407$

3 (a)

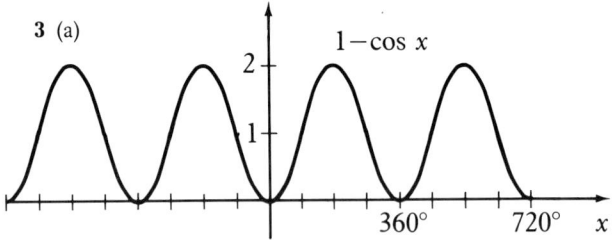

$1 - \cos x$

4 (a)

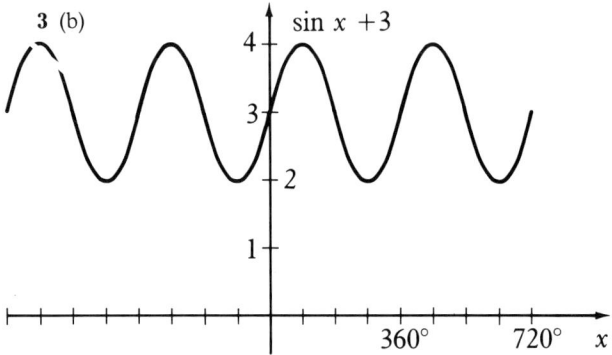

$3 \sin x$

period $= 360°$
amplitude $= 3$

3 (b)

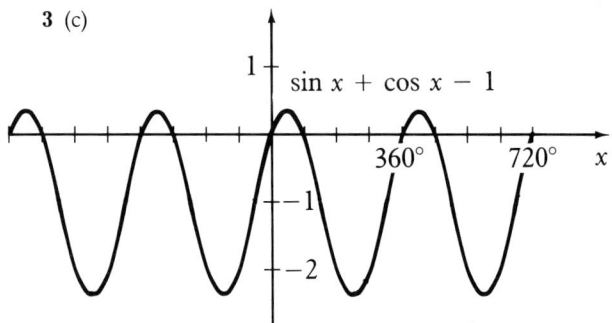

$\sin x + 3$

3 (c)

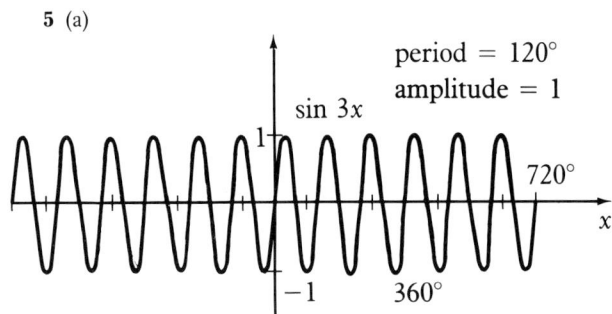

$\sin x + \cos x - 1$

4 (b)

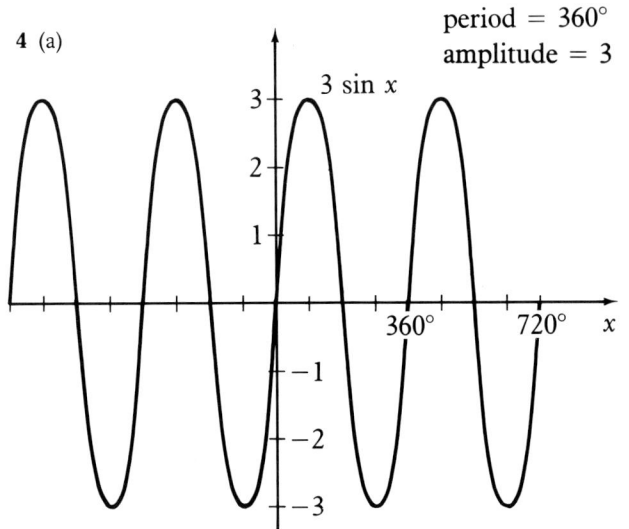

$0.5 \cos x$

period $= 360°$
amplitude $= 0.5$

5 (a)

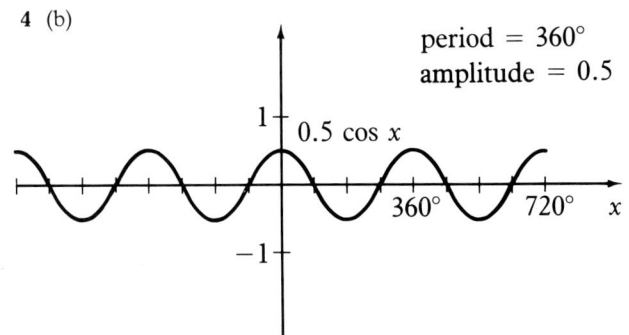

period $= 120°$
amplitude $= 1$

$\sin 3x$

5 (b)

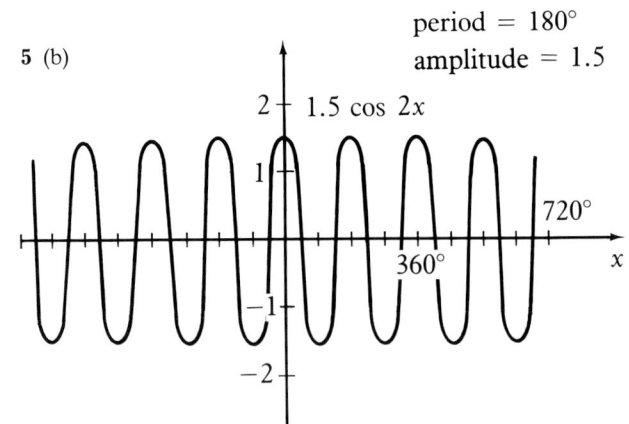

period $= 180°$
amplitude $= 1.5$

$1.5 \cos 2x$

5 (c)

period = $1080°$ $(3 \times 360°)$
amplitude = 1

cos $\frac{1}{3}$ x

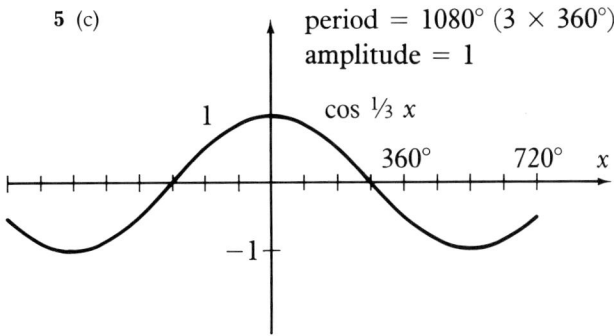

5 (d)

period = $720°$
amplitude = $\frac{1}{4}$

$\frac{1}{4}$ sin $\frac{1}{2}$ x

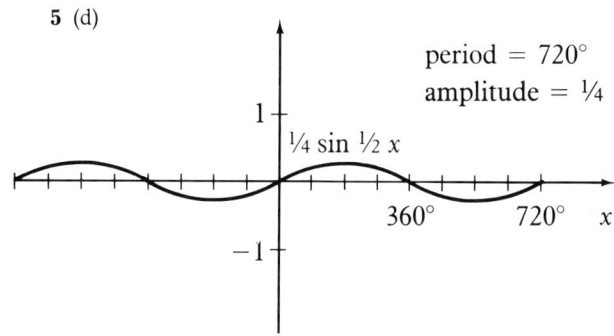

6 (a)

period = $360°$
amplitude = 1

sin $(x - 90°)$

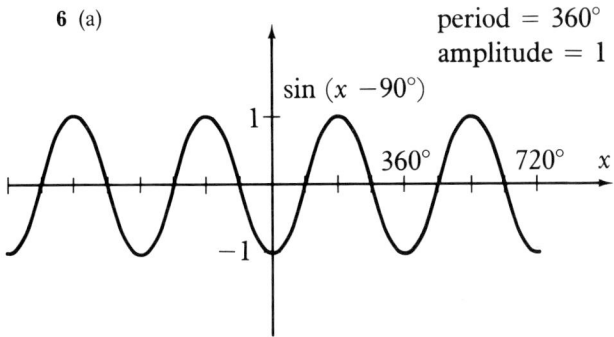

(b)

period = $360°$
amplitude = 1

cos $(x + 180°)$

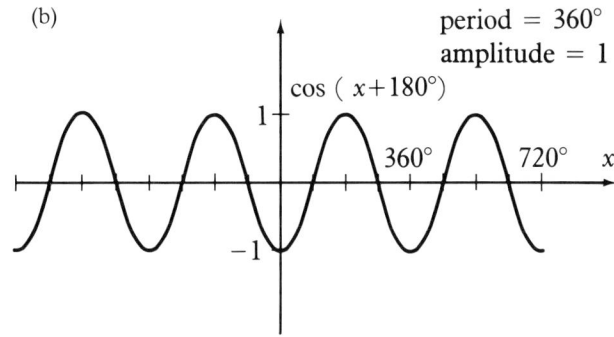

7 (a)

period = $180°$
amplitude = 2

$2 \cos^2 x$

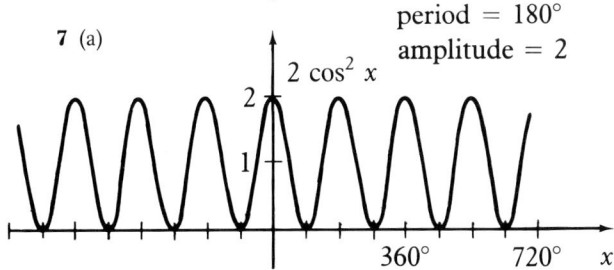

(b)

period = $180°$
amplitude = 0.5

$0.5 \sin^2 x$

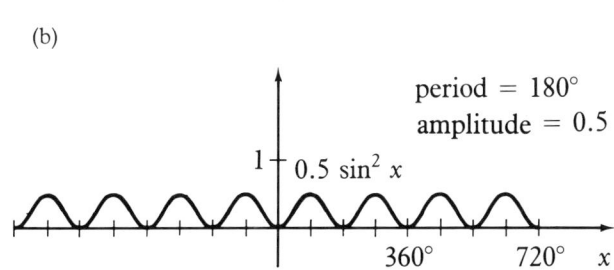

8 (a)

	tan 22° = 0·404	
tan 158° = − tan 22°	tan 158° = − 0·404	
tan 202° = tan 22°	tan 202° = 0·404	
tan 338° = − tan 22°	tan 338° = − 0·404	

(b)

tan 55° = 1·43	
tan 125° = − tan 55°	tan 125° = − 1·43
tan 235° = tan 55°	tan 235° = 1·43
tan 305° = − tan 55°	tan 305° = − 1·43

9

A	46°	52°	78°	102°	226°	308°
sin A	0·719	0·788	0·978	0·978	−0·719	−0·788
cos A	0·695	0·616	0·208	−0·208	−0·695	0·616
tan A	1·036	1·280	4·705	−4·705	1·036	−1·280

10 graph A: $f_2(x) = 2\sin\frac{1}{2}x$ graph B: $f_4(x) = \cos(x - 45)$ graph C: $f_1(x) = \frac{1}{2}\sin 2x$ graph D: $f_3(x) = \cos(x + 45)$

11 (a) $3\cos\frac{1}{2}x$; $a = 3$, $b = \frac{1}{2}$ (b) $2\sin(x + 90°)$; $a = 2$, $c = 90°$
   (c) $\frac{1}{2}\sin 3x$; $a = \frac{1}{2}$, $b = 3$ (d) $\frac{1}{2}\cos(x + 90°)$; $a = \frac{1}{2}$, $c = 90°$

Unit 8. Proportionality

1 (a) inverse proportion   (b) direct proportion
   (c) not in proportion   (d) approximately direct proportion

2 (a) $R \propto S$ (b) $t \propto \dfrac{1}{s}$ (c) $p \propto l$

3 (a) (i) $y = x^2 - 3$ (ii) $y = 2x - 3$ (iii) $y = \dfrac{2}{x}$ (iv) $y = \frac{1}{2}x$ (b) graph (iv)   (c) graph (iii)   **4** Tables (a) and (c)

5 (a) $q = 20p$ or $\dfrac{q}{p} = 20$ (b) 64   (c) 2·5   **6** (a) $p = 0·034d$ or $\dfrac{p}{d} = 0·034$ (b) £7·48

7 3·93 cm (to nearest 0·01 cm)

8 (a) 1·92 cm (to nearest 0·01 cm) (b) angle at centre of circle : length of arc

9 (a) $a = \dfrac{bc}{360}$ (b) $\dfrac{b\pi d}{360}$ **10** Yes; use formula in 9(a)

11 Angle at circumference $= \frac{1}{2}$ angle at centre; angle at centre $=$ length of arc   **12** $A \propto l^2$ **13** $V \propto l^3$

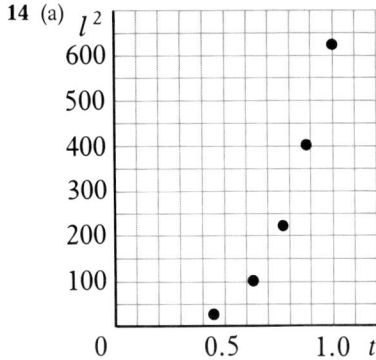

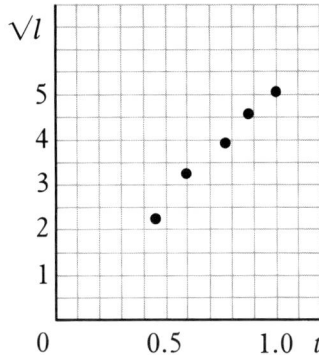

14 (a)

(b) $t \propto \sqrt{l}$ (c) $t = 0·2\sqrt{l}$

15 7·07 cm²   **16** 2·73 cm² (to 2 decimal places)   **17** $S = \dfrac{bA}{360}$

18 Table (b)   **19** (a) $xy = 28$ or $x = \dfrac{28}{y}$ (b) 14 (c) 2·8

20 (a) $p\sqrt{q} = 8$ or $p = \dfrac{8}{\sqrt{q}}$ (b) 1·6 (c) 4   **21** (a) $\frac{1}{2}$ (b) $\frac{1}{2}$

22 (a) $xy = 6$ or $x = \dfrac{6}{y}$ **22** (b)

23 (a) $cd = 2496$ or $c = \dfrac{2496}{d}$ (b) 32

24 (a) $p = 2qr$ (b) 70 (c) 5
25 (a) $A = 7lb$ (b) 38·08
26 (a) $x = 3yz^2$ (b) ± 6

27 (a) $l = \dfrac{4m}{n}$ (b) 8 (c) 8

28 (a) $a = \dfrac{b^2}{2c}$ (b) ± 6

29 Wind resistance is multiplied by 9.
30 (a) weight is doubled (b) weight is multiplied by 4

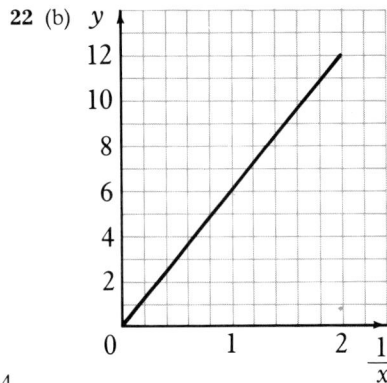

Revision—Proportionality

1 (a) inverse (b) neither (c) direct   **2** (a) neither (b) neither (c) direct (d) inverse

3 (a) $a = \frac{1}{2}b$ (b) $13\frac{1}{2}$ **4** (a) $pq = 35$ or $p = \dfrac{35}{q}$ (b) 5

5 (a) $l = \dfrac{3m}{n^2}$ (b) 5·25 (c) ± 5 **6** $V = \dfrac{kT}{p}$ (where k is a constant)   **7** (a) $W = 0·003\,125\ lr^2 = \dfrac{lr^2}{320}$ (b) 9 kg

8 (a) Water resistance is multiplied by 8. (b) Speed of boat is multiplied by $\sqrt[3]{2}$.

Unit 9. Trigonometry

1 (a) 4·91 (b) 10·26 (c) 12·08 (d) 27·64   **2** (a) 53·13° (b) 35·10° (c) 69·44°

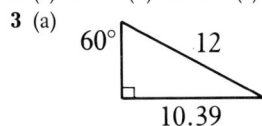

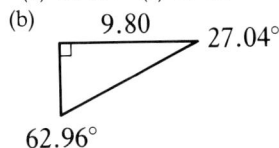

3 (a)

(b)

4 The triangle must contain a right angle. The length of two sides or the length of one side and the size of an angle (not the right angle).

5 9·56 cm   **6** 6·70 cm   **7** 5·85 cm   **8** 7·91 cm   **9** 81·87°   **10** 20·38°   **11** (i) 8·01 cm   (ii) 10·15 cm   (iii) 73°

12 (i) 3 pieces of information: either 2 angles and 1 side or 2 sides and 1 angle   (ii) sine   (iii) altitude

13 In questions 5–11, the angle given was opposite one of the known sides.

14 10·85 cm   **15** 9·58 cm   **16** 8·57 cm   **17** 10·99 cm   **18** 69·44°   **19** 58·96°

20 (a) 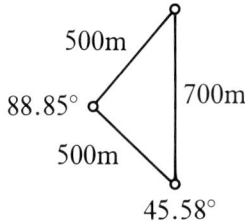 (b) 3840 m

45.58°

500m

88.85°

700m

500m

45.58°

21 6·17 cm   **22** 4·04 cm   **23** 28·44 cm

24 $p^2 = q^2 + r^2 - 2qr \cos P$ **25** $b^2 = a^2 + c^2 - 2ac \cos B$

26 7·21 cm   **27** 4·17 cm   **28** 12·80 cm

29 87·27°   **30** $\cos B = \dfrac{a^2 + c^2 - b^2}{2ac}$ **31** $\cos R = \dfrac{p^2 + q^2 - r^2}{2pq}$

32 62·80°   **33** 92·87°   **34** 82·82°   **35** 4·52 cm   **36** 100·29°   **37** 31·31°   **38** 10·52 cm   **39** 25·71 cm^2

40 (a) 15·43 cm^2 (b) 12·37 cm^2 (c) 39·01 cm^2 (d) 21·38 cm^2 (e) 43·52 cm^2 (f) 9·80 cm^2

41 (a) $AB \times AD \times \sin A$; i.e. product of two sides multiplied by sine of included angle

(b) $PQ \times QR \times \sin Q$ or $PS \times SR \times \sin S$

42 (a) 8·49 cm^2 (b) 45°   **43** (a) 14·42 cm^2 (b) 56·31°   (c) 22·59°

44 (a) 53·13°   (b) 41·82°   **45** 70·65°

46 (a) 45°, 225°   (b) 40°, 140°   (c) 230°, 310°   (d) 82°, 278°   (e) 30°, 150°   (f) 37°, 143°, 217°, 323°

(g) 150°, 210°

47 (a) $\tan 50°$ (b) $\sin A$ **48** 1   **49** (a) $4(\sin^2 A + \cos^2 A) = 4$ (b) $3\cos^2 A - 2$

50 $\tan A \ \sin A = \dfrac{\sin A}{\cos A} \ \sin A = \dfrac{\sin^2 A}{\cos A} = \dfrac{1 - \cos^2 A}{\cos A}$

Revision—Trigonometry

1 7·52 cm   **2** 20·27°   **3** 7·33 cm   **4** 100·29°   **5** 13·17 cm   **6** 30·25 cm^2 **7** 13·94 cm^2 **8** 15·16 cm^2

9 642·79 m   **10** (a) 7·07 cm   (b) 45°   **11** (a) 13·60 cm   (b) 40·60°   (c) 30·97°

12 (a) $(\sin A + \cos A)^2 = \sin^2 A + 2\sin A \ \cos A + \cos^2 A = 1 + 2\sin A \ \cos A$

(b) $\dfrac{\cos A}{\sin A} \times \tan A = \dfrac{\cos A}{\sin A} \times \dfrac{\sin A}{\cos A} = 1$

13 (a) 60°, 300°   (b) 11·31°, 191·31°   (c) 210°, 330°   (d) 82·87, 97·13°, 262·87°, 277·13°

Unit 10.   Transformations and matrices

1 If $P(x, y)$ is reflected in the line $y = -x$ its image is $P'(-y, -x)$

If P' is the reflection of a point P in the line $y = -x$, the x-coordinate of P' is the negative of the y-coordinate of P and the y-coordinate of P' is the negative of the x-coordinate of P.

If (x', y') is the reflection of (x, y) in the line $y = -x$; $x' = -y$, $y' = -x$

2 (a) $a = 135°$ $2a = 270°$ $\sin 2a = -1$ $\cos 2a = 0$

$x' = (\cos 2a)x + (\sin 2a)y = 0x + (-1)y = -y$ $y' = (\sin 2a)x - (\cos 2a)y = (-1)x - 0y = -x$

These results are the same as those in Try This 1.

(b) $a = 90°$ $2a = 180°$ $\sin 2a = 0$ $\cos 2a = -1$

$x' = (\cos 2a)x + (\sin 2a)y = -1x + 0y = -x$ $y' = (\sin 2a)x - (\cos 2a)y = 0x - (-1)y = y$

3 $x' = x$ $y' = 2j - y$

4 (a) $(50, -10)$ (b) $(40, 20)$ (c) $(-20, 56)$ (d) $(10, -40)$ (e) $(30, 50)$ (f) $(-18, 10)$

5 half-turn $x' = -x$, $y' = -y$ quarter-turn anticlockwise $x' = -y$, $y' = x$ quarter-turn clockwise $x' = y$, $y' = -x$

6 (a) $a = 90°$; $x' = -y$, $y' = x$ (b) $a = -90°$; $x' = y$, $y' = -x$

7 (a) $(1, -5)$ (b) $(2, 4)$ (c) $(-5, 2)$ (d) $(-4, 1)$ (e) $(-5, -3)$ (f) $(1, -1)$

8 (a) $(0,2)$ (b) $(2,2)$ (c) $(-1,-8)$ (d) $(2,3)$ (e) $(-4,6)$ (f) $(-4,6)$
9 (a) $(-3,15)$ (b) $(2,-1)$ (c) $(4,10)$ (d) $(-0\cdot4,0\cdot1)$ (e) $(-30,50)$ (f) $(-0\cdot5,-0\cdot5)$
10 $x'=2x, \ y'=y$

$(1,1) \longrightarrow (2,1)$
$(1,3) \longrightarrow (2,3)$
$(4,1) \longrightarrow (8,1)$
$(4,3) \longrightarrow (8,3)$

This transformation is a dilatation in one direction only — a 'one way stretch' — (in the x-direction in this case)

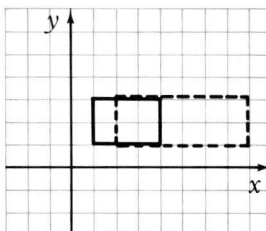

11 (a) $\begin{pmatrix} 0 & -1 \\ -1 & 0 \end{pmatrix}$ (b) $\begin{pmatrix} 1 & 0 \\ 0 & -1 \end{pmatrix}$ **12** (a) $\begin{pmatrix} 0 & -1 \\ 1 & 0 \end{pmatrix}$ (b) $\begin{pmatrix} 0 & 1 \\ -1 & 0 \end{pmatrix}$

13 (a) Inverse matrix is $\begin{pmatrix} 0 & 1 \\ -1 & 0 \end{pmatrix}$ which is the matrix of the transformation *rotation about O through* $-90°$ (quarter-turn clockwise). We would expect the inverse of an anticlockwise turn to be a clockwise turn.

(b) Inverse matrix is $\begin{pmatrix} 0 & -1 \\ 1 & 0 \end{pmatrix}$ which is the matrix of the transformation *rotation about O through* $90°$ (quarter-turn anticlockwise). We would expect the inverse of a clockwise turn to be an anticlockwise turn.

14 (a) a half-turn followed by a quarter-turn anticlockwise

$\begin{pmatrix} 0 & -1 \\ 1 & 0 \end{pmatrix} \begin{pmatrix} -1 & 0 \\ 0 & -1 \end{pmatrix} = \begin{pmatrix} 0 & 1 \\ -1 & 0 \end{pmatrix}$ a quarter-turn clockwise

(b) a quarter-turn anti-clockwise followed by a quarter-turn clockwise

$\begin{pmatrix} 0 & 1 \\ -1 & 0 \end{pmatrix} \begin{pmatrix} 0 & -1 \\ 1 & 0 \end{pmatrix} = \begin{pmatrix} 1 & 0 \\ 0 & 1 \end{pmatrix}$ a 'no change' transformation (identity transformation)

(c) reflection in line $y=x$ followed by reflection in line $y=-x$

$\begin{pmatrix} 0 & -1 \\ -1 & 0 \end{pmatrix} \begin{pmatrix} 0 & 1 \\ 1 & 0 \end{pmatrix} = \begin{pmatrix} -1 & 0 \\ 0 & -1 \end{pmatrix}$ a half-turn

(d) reflection in the x-axis followed by reflection in the y-axis

$\begin{pmatrix} -1 & 0 \\ 0 & 1 \end{pmatrix} \begin{pmatrix} 1 & 0 \\ 0 & -1 \end{pmatrix} = \begin{pmatrix} -1 & 0 \\ 0 & -1 \end{pmatrix}$ a half-turn

(e) reflection in the line $y=x$ followed by reflection in the x-axis

$\begin{pmatrix} 1 & 0 \\ 0 & -1 \end{pmatrix} \begin{pmatrix} 0 & 1 \\ 1 & 0 \end{pmatrix} = \begin{pmatrix} 0 & 1 \\ -1 & 0 \end{pmatrix}$ a quarter-turn clockwise

(f) reflection in the line $y=x$ followed by reflection in the y-axis

$\begin{pmatrix} -1 & 0 \\ 0 & 1 \end{pmatrix} \begin{pmatrix} 0 & 1 \\ 1 & 0 \end{pmatrix} = \begin{pmatrix} 0 & -1 \\ 1 & 0 \end{pmatrix}$ a quarter-turn anticlockwise

(g) reflection in the line $y=-x$ followed by reflection in the x-axis

$\begin{pmatrix} 1 & 0 \\ 0 & -1 \end{pmatrix} \begin{pmatrix} 0 & -1 \\ -1 & 0 \end{pmatrix} = \begin{pmatrix} 0 & -1 \\ 1 & 0 \end{pmatrix}$ a quarter-turn anticlockwise

(h) reflection in the line $y=-x$ followed by reflection in the y-axis

$\begin{pmatrix} -1 & 0 \\ 0 & 1 \end{pmatrix} \begin{pmatrix} 0 & -1 \\ -1 & 0 \end{pmatrix} = \begin{pmatrix} 0 & 1 \\ -1 & 0 \end{pmatrix}$ a quarter-turn clockwise

Revision—Transformations and matrices

1 (a) A'(4,1), B'(2,-3), C'(-5,0) (b) A'(-4,-1), B'(-2,3), C'(5,0)
(c) A'(1,-4), B'(-3,-2), C'(0,5) (d) A'(-1,4), B'(3,2), C'(0,-5)
(e) A'(3,4), B'(7,2), C'(4,-5) (f) A'(1,-10), B'(-3,-8), C'(0,-1)
(g) A'(-1,-4), B'(3,-2), C'(0,5) (h) A'(-4,1), B'(-2,-3), C'(5,0)
(i) A'(4,-1), B'(2,3), C'(-5,0) (j) A'(2,2), B'(-2,0), C'(1,-7) (k) A'(2,8), B'(-6,4), C'(0,-10)

2
(a) $\begin{pmatrix} 0 & 1 \\ -1 & 0 \end{pmatrix} \begin{pmatrix} -1 & 0 \\ 0 & -1 \end{pmatrix} = \begin{pmatrix} 0 & -1 \\ 1 & 0 \end{pmatrix}$ quarter-turn anticlockwise

(b) $\begin{pmatrix} -1 & 0 \\ 0 & -1 \end{pmatrix} \begin{pmatrix} 0 & -1 \\ 1 & 0 \end{pmatrix} = \begin{pmatrix} 0 & 1 \\ -1 & 0 \end{pmatrix}$ quarter-turn clockwise

(c) $\begin{pmatrix} 0 & 1 \\ 1 & 0 \end{pmatrix} \begin{pmatrix} 0 & -1 \\ -1 & 0 \end{pmatrix} = \begin{pmatrix} -1 & 0 \\ 0 & -1 \end{pmatrix}$ half-turn

(d) $\begin{pmatrix} 1 & 0 \\ 0 & -1 \end{pmatrix} \begin{pmatrix} -1 & 0 \\ 0 & 1 \end{pmatrix} = \begin{pmatrix} -1 & 0 \\ 0 & -1 \end{pmatrix}$ half-turn

(e) $\begin{pmatrix} 0 & 1 \\ 1 & 0 \end{pmatrix} \begin{pmatrix} 1 & 0 \\ 0 & -1 \end{pmatrix} = \begin{pmatrix} 0 & -1 \\ 1 & 0 \end{pmatrix}$ quarter-turn anticlockwise

(f) $\begin{pmatrix} 0 & 1 \\ 1 & 0 \end{pmatrix} \begin{pmatrix} -1 & 0 \\ 0 & 1 \end{pmatrix} = \begin{pmatrix} 0 & 1 \\ -1 & 0 \end{pmatrix}$ quarter-turn clockwise

(g) $\begin{pmatrix} 0 & -1 \\ -1 & 0 \end{pmatrix} \begin{pmatrix} 1 & 0 \\ 0 & -1 \end{pmatrix} = \begin{pmatrix} 0 & 1 \\ -1 & 0 \end{pmatrix}$ quarter-turn clockwise

(h) $\begin{pmatrix} 0 & -1 \\ -1 & 0 \end{pmatrix} \begin{pmatrix} -1 & 0 \\ 0 & 1 \end{pmatrix} = \begin{pmatrix} 0 & -1 \\ 1 & 0 \end{pmatrix}$ quarter-turn anticlockwise

Unit 11.   Statistics

1

16	5 7 9 9 9
17	1 1 3 3 4 5 9
18	0 0 1 1 1 2 3 5

2

Class X		Class Y	
0		0	
1	2	1	2 3
2	5 9	2	
3	8	3	4
4	5 9	4	4 8
5	5 7 8	5	3 9
6	5 7 8	6	2 3 6
7	5 8 9	7	0 0 1 2 4
8	2 3 7	8	0 1 2
9	1 9	9	0 1

3 mean = 61·2 marks

4 X mean = 62·1p
    Y mean = 61·75p

5

Mark	Mid-point (M)	Frequency (f)	f × M
30–39	34·5	3	103·5
40–49	44·5	6	267
50–59	54·5	7	381·5
60–69	64·5	4	258
70–79	74·5	5	372·5
80–89	84·5	3	253·5
90–99	94·5	2	189
Totals		30	1825

$$\text{Mean} = \frac{1825}{30} = 61$$

6 CLASS X

Amount	Mid-point (M)	Frequency (f)	f × M
0–19	9·5	1	9·5
20–39	29·5	3	88·5
40–59	49·5	5	247·5
60–79	69·5	6	417
80–99	89·5	5	447·5
Totals		20	1210

$$\text{Mean} = \frac{1210}{20} = 60·5$$

CLASS Y

Amount	Mid-point (M)	Frequency (f)	f × M
0–19	9·5	2	19
20–39	29·5	1	29·5
40–59	49·5	4	198
60–79	69·5	8	556
80–99	89·5	5	447·5
Totals		20	1250

$$\text{Mean} = \frac{1250}{20} = 62·5$$

7 The 15th mark is 56; the 16th mark is 57; the median is 56·5

8 CLASS X: the 10th amount is 65p; the 11th amount is 67p; the median is 66p
    CLASS Y: the 10th amount is 66p; the 11th amount is 70p; the median is 68p

9 Cumulative frequencies: 3 9 16 20 25 28 30

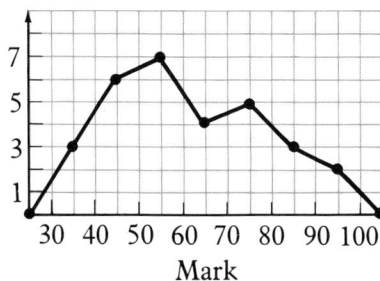

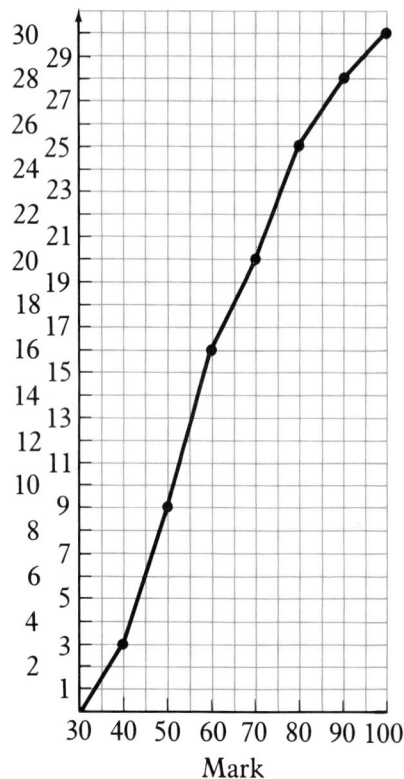

10 Cumulative frequencies—CLASS X: 1 4 9 15 20
CLASS Y: 2 3 7 15 20

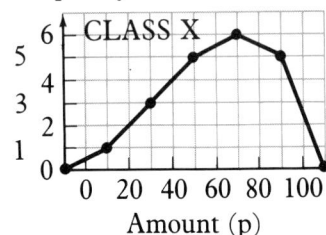

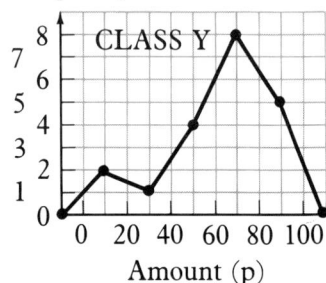

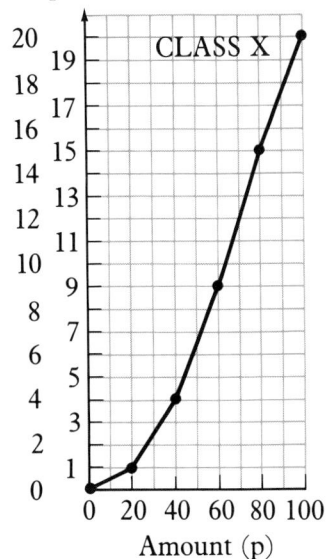

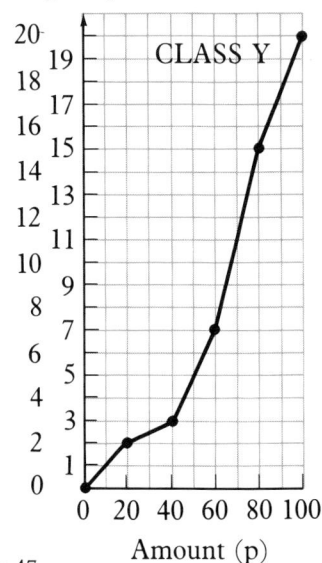

11 8th mark is 47; lower quartile, Q_1, is 47.
23rd mark is 75; upper quartile, Q_3, is 75.   interquartile range $= 75 - 47 = 28$

12 CLASS X: the 5th amount is 45p; the 6th amount is 49p; Q_1 is 47p. The 15th amount is 79p; the 16th amount is 82p; Q_3 is 80·5p. Semi-interquartile range $(Q_3 - Q_1)/2 = (80·5 - 47)/2 = 16·75$p
CLASS Y: the 5th amount is 48p; the 6th amount is 53p; Q_1 is 50·5p. The 15th amount is 74p; the 16th amount is 80p; Q_3 is 77p. Semi-interquartile range $(Q_3 - Q_1)/2 = (77 - 50·5)/2 = 13·25$p

13 (a) mode $= 54$ (b) modal class is 50-59

Revision—Statistics

1 (a) Mean 5; median 5; mode 3 (b) Mean 4·5; median 4; mode 2 (c) Mean 2·7 m; median 2·6 m; mode 2·6 m
2 5 **3** mean 50·3; median 52; mode 52
4 (a) mean 1·59 m; median 1·6 m; mode 1·6 m (b) upper 1·7 m; lower 1·45 m **5** (a) 74 (b) 74
6 (a) 113 miles (b) upper 127 miles; lower 102 miles (c) 25